This page is intentionally left blank

Vertical Farming
A Guide for Growing Minds

Maryna Kuzmenko

Petiole Pro R&D Lab

London, United Kingdom

ISBN: 9798335398282
Vertical Farming: A Guide for Growing Minds
By Maryna Kuzmenko

Copyright © 2024 Maryna Kuzmenko. All rights reserved.

Published by Petiole R&D Lab, Petiole LTD, 85 Great Portland Street (First floor), London, W1W 7LT, United Kingdom

Available for purchase on Amazon

Petiole Pro R&D Lab books may be purchased for educational, business, or sales promotional use. Online editions are also available for most titles (*https://petiolepro.com*). For more information, contact us: *support@petioleapp.com*

Editor: Andrii Seleznov

Revision History for the First Edition

2024-08-08 First Release

See *http://petiolepro.com/vertical-farming-a-guide-for-growing-minds* for release details.

The Petiole logo is a registered trademark of Petiole LTD.
Vertical Farming: A Guide for Growing Minds, the cover image and related trade dress are patented by Petiole LTD.

The publisher and authors have made every effort to ensure that the information and instructions contained in this work are accurate. However, they disclaim all responsibility for errors or omissions, including, without limitation, any responsibility for damages resulting from the use of or reliance on this work. The use of the information and instructions contained in this work is **at your own risk**. If any part of this book or technology described in this work are subject to open source licences or the intellectual property rights of others, it is **your responsibility** to ensure that your use complies with such licences and/or rights.

No part of this publication may be reproduced, distributed, or transmitted in any form or by any means, including photocopying, recording, or other electronic or mechanical methods, without the prior written permission of the publisher, except in the case of brief quotations embodied in critical reviews and certain other noncommercial uses permitted by copyright law. For permission requests, write to the publisher at the address above.

Table of Contents

Preface..5
1. Introduction to Vertical Farming.. 6
    Definition and Concept... 7
    Historical Context and Pioneers.. 12
    How Vertical Farming Has Changed... 31
    Global Relevance and Potential.. 33
2. The Science Behind Vertical Farming... 36
    Plant Biology Basics... 37
    Comparing Indoor and Outdoor Vertical Farming......................... 56
    Soilless and Substrate-based Growing Methods......................... 71
3. Vertical Farming Layout.. 82
    Crop Selection and Crop Health... 83
    Environment.. 123
    AI and Internet of Things (IoT).. 151
4. Vertical Farming Economics... 177
5. Design and Construction.. 188
    Architectural Considerations... 189
    Structural and Environmental Design Elements......................... 192
    Design and Integration with Existing Buildings...........................205
6. Advancement in Vertical Farming... 212
    Breeding and Genetics..213
    Precise Nutrient Delivery...220
    Optimization of Environmental Control..233
    Beneficial Microorganisms.. 244
    Artificial Pollination.. 254
7. Sustainability and Resource Management................................... 259
    Integrated Rainwater Harvesting...260
    Renewable Energy Integration..263
    Life Cycle Assessment.. 271
8. Case Studies of Successful Vertical and Urban Farms................ 278
What's next?.. 286
Bibliography...287
About the Author.. 294
About the Publisher..295

Preface

More people. Less land. Climate change. These are the challenges that have pushed traditional agricultural methods to their limits in meeting the rising demand for fresh produce. Conventional farming practices are now associated with widespread deforestation, excessive water usage, and significant greenhouse gas emissions, all contributing to the global environmental crisis.

In the face of these challenges, **vertical farming has emerged as a beacon of hope and promise.** This innovative approach, which involves growing crops in stacked layers within controlled environments, offers numerous advantages over traditional farming. However, for wider adoption, **vertical farming needs more voices to spread awareness about this revolutionary agricultural practice.**

That's why this book was born – to explore the multifaceted world of vertical farming through scientific and technological advancements from around the globe. Based on **a collection of hand-picked scientific reports**, this book provides practical insights for entrepreneurs, policymakers, and researchers interested in the vertical farming sector.

While just a starting point, it aims to inspire further exploration and development in vertical farming, showing the way for **sustainable food systems that benefit both humanity and the environment**.

Welcome to the future of farming.

1. Introduction to Vertical Farming

This chapter, *Introduction to Vertical Farming,* serves as a foundational overview of the concept, its evolution, and its significance in modern agriculture. Here's a brief explanation of its contents and their importance:

Definition and Concept provides a clear explanation of vertical farming, outlining its **core principles and distinguishing features**. It's important to read it to establish a common understanding for readers new to the topic.

Historical Context and Pioneers explores the **origins of vertical farming** and highlights **key figures** who contributed to its development. It helps readers appreciate how the concept has evolved and who played significant roles in shaping it.

How Vertical Farming Has Changed traces the **technological and methodological advancements** in vertical farming over time. Understanding this evolution is important for grasping current practices and anticipating future trends.

Global Relevance and Potential discusses the **worldwide applications** and **future prospects** of vertical farming. It's essential for understanding the broader impact and possibilities of this agricultural approach in addressing global challenges like food security and urban sustainability.

Definition and Concept

Vertical farming represents a revolutionary approach to agriculture that challenges traditional ideas about growing food. At its core, **vertical farming** is the practice of growing crops in **vertically stacked layers**, often incorporating **controlled-environment agriculture technology**. This innovative method optimises plant growth, maximises space efficiency, and allows for year-round cultivation regardless of external weather conditions.

> The concept of vertical farming is built on **seven key pillars**: vertical space utilisation, controlled environment, soilless growing techniques, artificial lighting, technology integration, urban agriculture and sustainability focus.

Vertical space utilisation
Unlike conventional horizontal farming, vertical farms take advantage of vertical space, allowing for significantly higher crop yields per square foot of land area. This is achieved through the use of stacked growing systems, often reaching several stories high in indoor facilities.

Controlled environment
Vertical farms typically operate in enclosed spaces where environmental factors such as temperature, humidity, light, and CO_2 levels can be precisely controlled. This creates the optimal conditions for plants, leading to faster growth and higher yields.

Soilless growing techniques
Most vertical farms employ hydroponic, aeroponic, or aquaponic systems. These methods deliver nutrients directly to plant roots, removing the need for soil and reducing water usage by up to 95% compared to traditional farming [Carotti *et al.*, 2023].

Artificial lighting
In indoor vertical farms, plants receive their light from artificial sources, typically light emitting diode (LED) grow lights. These lights can be adjusted to give the optimal light for each type of plant, helping them grow better and faster.

Technology integration
Advanced technologies such as sensors, automation systems, and data analytics play an important role in monitoring and optimising plant growth, resource use, and overall farm efficiency.

Urban agriculture

While not exclusive to urban areas, vertical farming is particularly well-suited to cities, where it can bring food production closer to consumers, reducing transportation costs and carbon emissions associated with long-distance food shipping.

Sustainability focus
By design, vertical farming aims to address several environmental concerns associated with traditional agriculture, including land use, water consumption, and pesticide use.

These pillars will be explored in more detail in the following chapters.

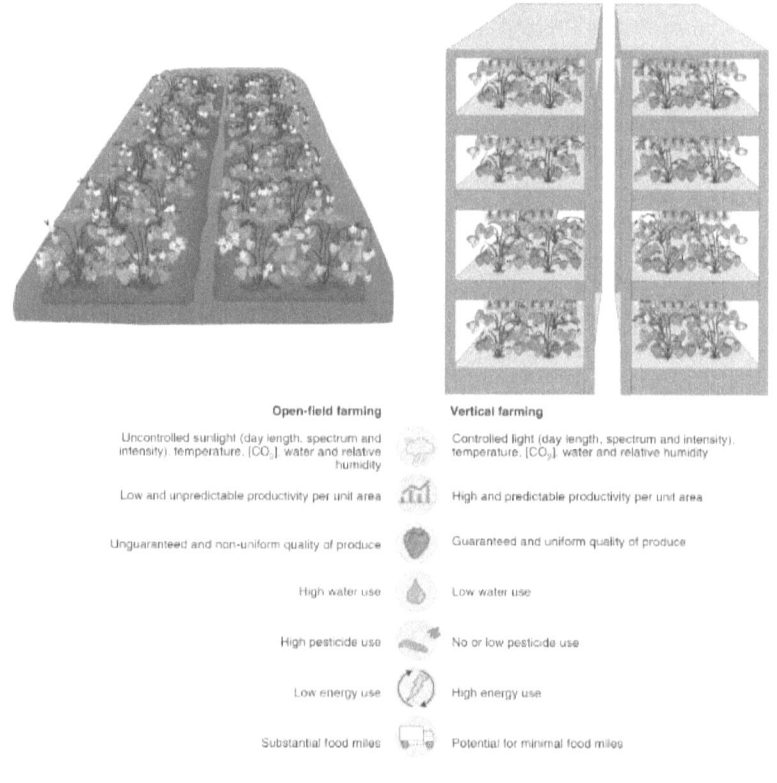

Comparison of open field farming and vertical farming.
Source: van Delden *et al*, 2021

Vertical farming is **more than just a new way to grow food**. It's a big change in how we think about farming, cities, and where our food comes from.

The United Nations says that by the year 2050:

- There will be about **9 billion people** in the world
- **7 out of every 10** people **will live in cities** [FAO, 2009].

This new way of farming could help feed more people and make our food healthier and fresher.

Vertical farming helps us think about:

1. How to **grow food in small spaces** in cities
2. How to **make cities better places to live**
3. How to **bring farms and people in cities closer** together

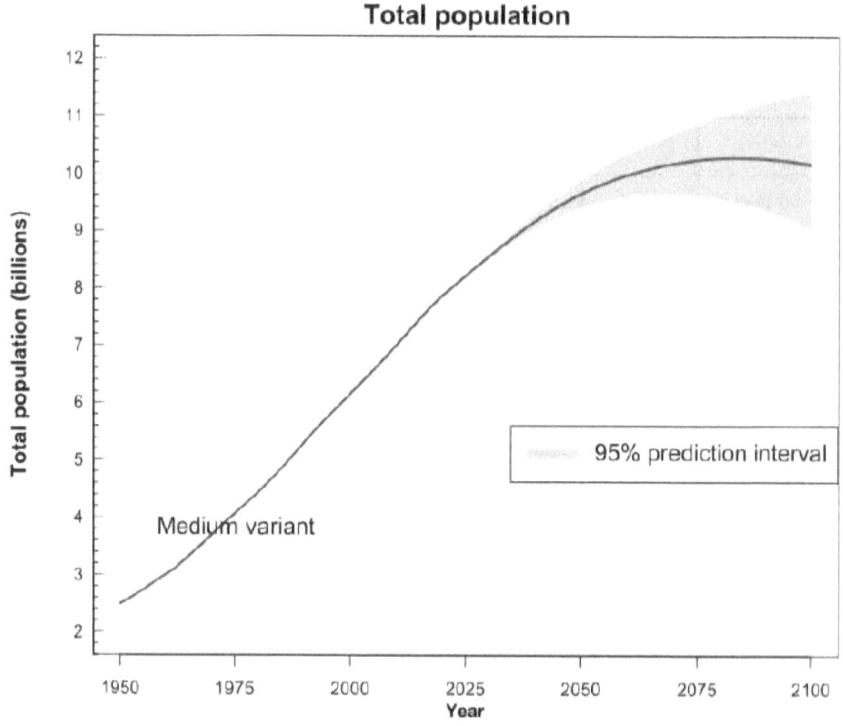

The UN projects significant population growth from 2009 to 2050, reaching 9.1 billion people. Source: United Nations, Population Division / *World Population Prospects 2024*.

As more people move to cities and climate change makes traditional farming harder, vertical farming offers a **way to grow food locally and sustainably**. By using new technologies and combining them with natural processes, **vertical farming blends agriculture, engineering, and environmental science**.

This innovative method can make our food systems stronger and more sustainable, helping to feed the growing global population in the future.

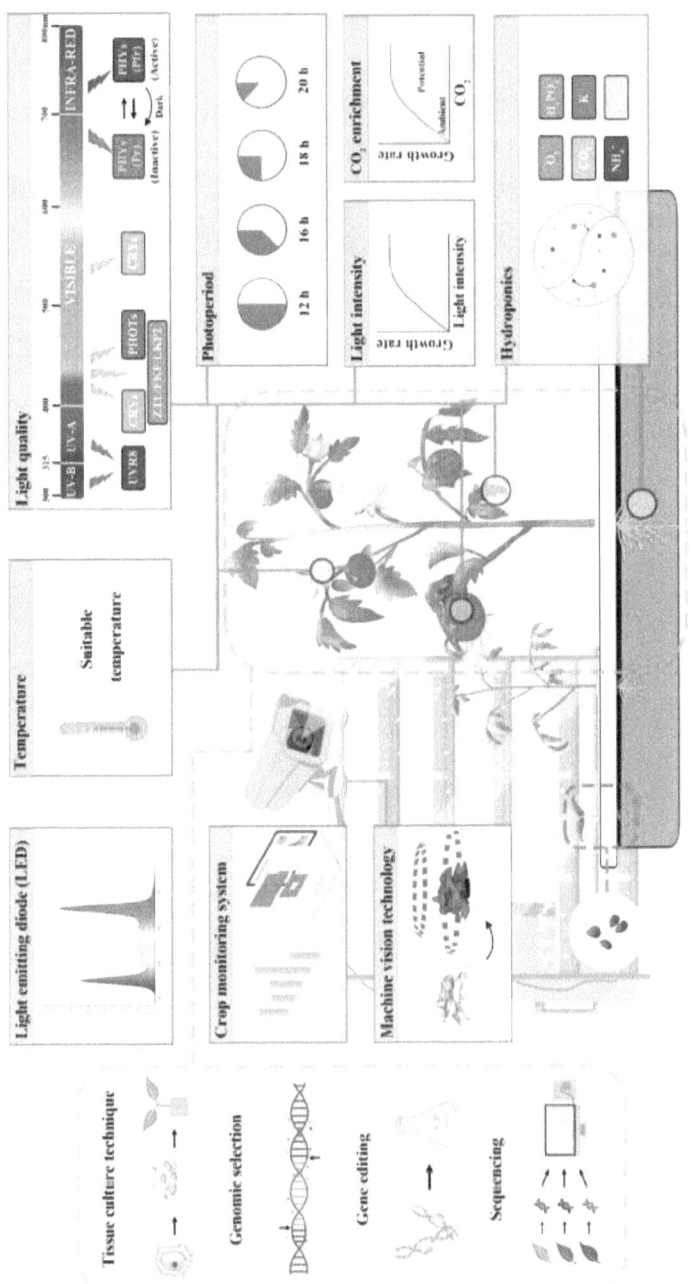

Generalised view of vertical garden versus horizontal garden. Horizontal gardening involves crops grown in soil as a single bed system, but vertical gardening increases the number of growth beds per land plot to accommodate more plants and potentially increase yield per plot. Source: Mishra *et al.*, 2021

Historical Context and Pioneers

Vertical farming, often seen as a modern innovation, actually **has deep historical roots** and has evolved significantly over time due to technological advancements and urban population growth.

> The earliest ideas of the vertical farming concept and **space-efficient approach to farming** include the Hanging Gardens of Babylon, Aztec "chinampas", mediaeval European castle gardens, Persian qanats, and terraced farming methods like the Incan agricultural terraces and the Banaue Rice Terraces in the Philippines.

All these early practices of vertical farming and space-efficient farming share the common principle of **efficiently using limited space to improve agricultural output.**

They each employed **innovative methods for their time**, utilising vertical or tiered structures, inventing efficient irrigation systems, and strategically placing crops to enhance food production in challenging environments.

Hanging Gardens of Babylon, constructed c. 8th–6th century BCE. Artist's re-creation. Source: *Britannica* / Chronicle / Alamy

Remains of the Hanging Gardens of Babylon. Source: *Google Maps* / Hallow Hama Radha

Chinampas and canals, 1912. Source: *Wikipedia*

Modern chinampas as a part of land reclamation project. Source: *BBC Travel* / Arca Tierra

The kitchen garden at Alfriston Clergy House, East Sussex. Source: *National Trust Images* / James Dobson

Aerial View of the Persian Qanat, Jupar, Bagh-e Shahzadeh (Mahan). Source: *UNESCO* / S.H. Rashedi

The Terraces at Pisac in the sacred valley of the Incas, in Peru. Source: *National Geographic Education*

The Banaue Rice Terraces were carved into the mountains of Banaue, Ifugao, in the Philippines, by the ancestors of the indigenous people. Source: *Wikimedia* / Ranieljosecastaneda

The history of vertical farming lies at the **intersection of horticulture, architecture, engineering, and other fields**, with roots tracing back to early experiments in plant cultivation and nutrient delivery.

British scientist **John Woodward** (1665-1728) conducted a groundbreaking experiment in 1699, demonstrating that **plants could grow in less-pure water with the addition of soil**, laying the foundation for understanding plant nutrition and paving the way for soilless cultivation techniques.

Building on this knowledge, German scientist **Wilhelm Knop** (1817-1891) and Russian scientist **Kliment Timiryazev** (1843-1920) independently conducted experiments in the late 19th century that established the **principles of hydroponics**, demonstrating that plants could grow in nutrient-rich water solutions without soil, a concept fundamental to modern vertical farming practices.

The idea of vertical farming reappeared later in 1915 when American scientist **Gilbert Ellis Bailey** (1852 – 1924) published his book "Vertical Farming". Bailey's vision was different from today's methods.

He imagined using vertical layers of soil for farming, instead of the stacked shelves or hydroponic systems used in modern vertical farming.

His book was considered **groundbreaking in soil science** and agriculture during its time. However, considering that vertical farming focuses more on soilless growing and substrates, the **practical applications of the book are limited.**

Cover of the book "Vertical Farming" by Gilbert Ellis Bailey. Source: *Niagara Pen Centre*

In the 1930s, significant progress was made with the development of hydroponics. An American scientist, **William Frederick Gericke** (1882 – 1970) began experimenting with soilless farming methods, demonstrating that plants could be successfully grown in nutrient-rich water solution

> Gericke, often considered the **father of hydroponics**, showcased large-scale hydroponic farming to the public, proving its viability.

This breakthrough laid the **groundwork for the advanced hydroponic systems** widely used in vertical farming today, enabling efficient nutrient delivery and water conservation.

Gericke, standing on a ladder, harvests one of his tall tomato plants grown in hydroponics. His wife, Mrs. Gericke stands below with some of the harvest. Source: The website of the City of Winston-Salem, NC

Gericke grew tobacco plants that reached 15 feet tall (4.57 metres).
Source: Institute of Simplified Hydroponics

Othmar Ruthner (1912 – 1991) was an Austrian botanist and engineer who played a pioneering role in the development of vertical farming concepts.

In the 1960s, Ruthner designed and patented one of the **first modern vertical growing systems**. His "Ruthner Tower" was a cylindrical greenhouse with a rotating mechanism that moved plants on trays through different levels of the structure.

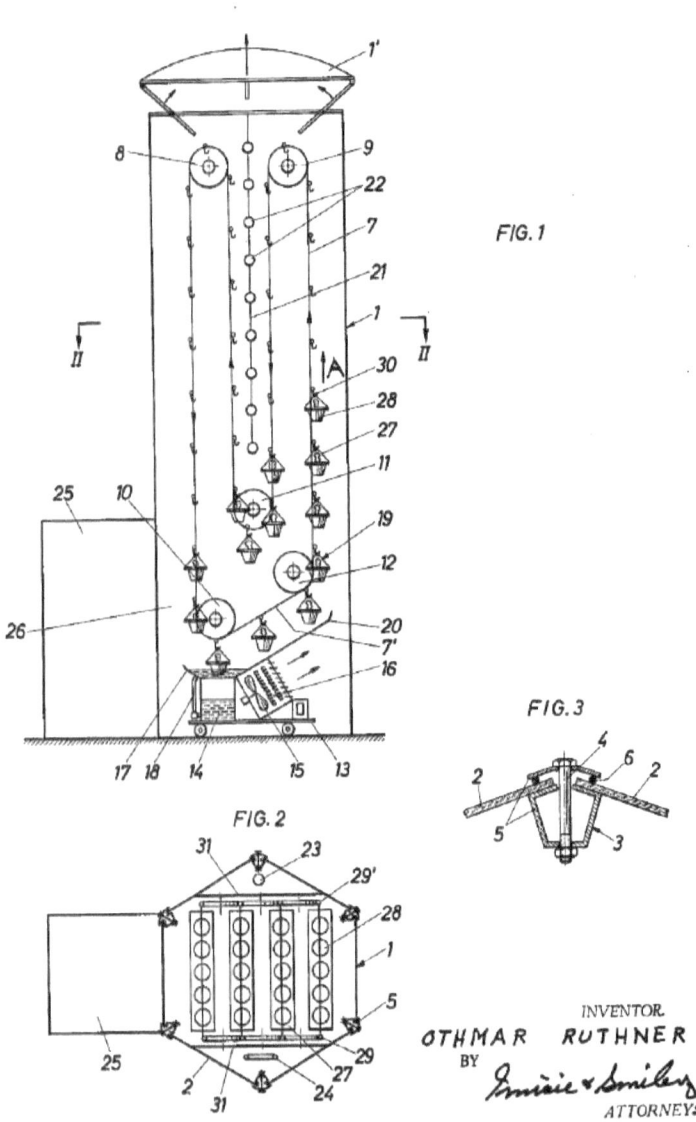

Illustration from the Ruthner patent for the tower greenhouse, 1966.
Source: *Google Patents* / US3254448A

"Culture tube" with cucumbers in the "rotating greenhouse". From the dissertation of Othmar Ruthner (1972). Source: Christian Hlavac

This design allowed for **efficient use of space and controlled exposure to light and nutrients**. Ruthner's system was showcased at various international exhibitions, including the 1964 Vienna International Garden Show, where it garnered significant attention.

Ruthner was also known for his work on **hydroponic systems**, which played a crucial role in his vertical farming designs.

He was a proponent of **using nutrient solutions instead of soil** to grow plants, which allowed for greater control over the growing environment and improved plant health.

(A) A view of the gardening tower at the WIG 64 exhibition in Vienna, 1964. (B) A view of the gardening tower, 1963. Source: Österreichisches Gartenbaumuseum / Kleszcz et al., 2020

Gagric Davtyan (1909 – 1980), an Armenian / USSR agronomist, made substantial contributions to hydroponics that later influenced vertical farming concepts. Davtyan's work showcased the potential for growing crops in controlled environments with minimal space requirements.

> Davtyan's research on nutrient solutions and plant growth in artificial conditions provided valuable insights that would later be **applied in vertical farming systems**.

Furthermore, Davtyan's success in cultivating plants in adverse conditions proved that with the right technology, **food could be grown almost anywhere** – a concept central to modern vertical farming. Davtyan was one of the founders of the Institute of Hydroponics Problems in Yerevan in 1947, which was later named after him.

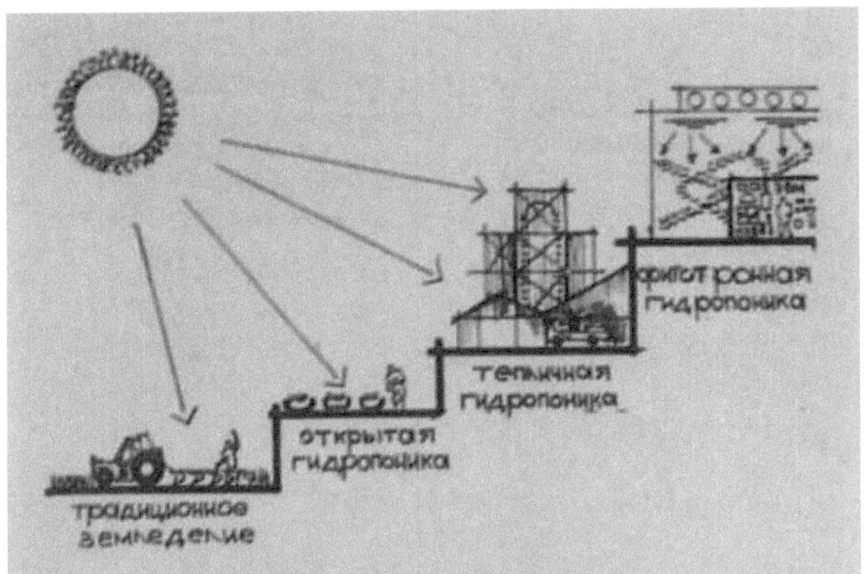

Stages of development of industrial horticulture without soil (traditional arable farming, open hydroponics, greenhouse hydroponics, phytotronical hydroponics). Source: Pan Armenian Digital Library / Davtyan Gagric

The modern era of vertical farming was profoundly influenced by American scientist **Dickson Despommier** (born 1940), a professor at Columbia University, who in 1999 introduced a concept that would reshape urban agriculture. Despommier, along with his graduate students, developed the **idea of the "vertical farm"** as a solution to growing food shortages and the environmental impact of traditional farming.

His concept went beyond mere food production; it encompassed a **holistic approach to urban sustainability**, addressing issues such as water conservation, energy efficiency, and reduction of transportation costs. His 2010 book, "The Vertical Farm: Feeding the World in the 21st Century," popularised the idea globally, inspiring architects, engineers, and entrepreneurs to pursue vertical farming projects.

> The core forces driving changes in architectural concepts have been the **increasing demand for sustainable urban agriculture** and the **availability of new materials and technologies**. These factors have led to innovations that make better use of space, use fewer resources, boost food security, and improve energy efficiency and durability in vertical farming designs.

Over the years, there have been several innovative **architectural concepts in vertical farming** that have significantly evolved, reshaping the approach to space utilisation and sustainability.

Le Corbusier's Immeubles-Villas (1922)

Multi storey buildings where people could live in "sky houses" stacked on top of each other. Each home had its own garden, bringing nature into city life.

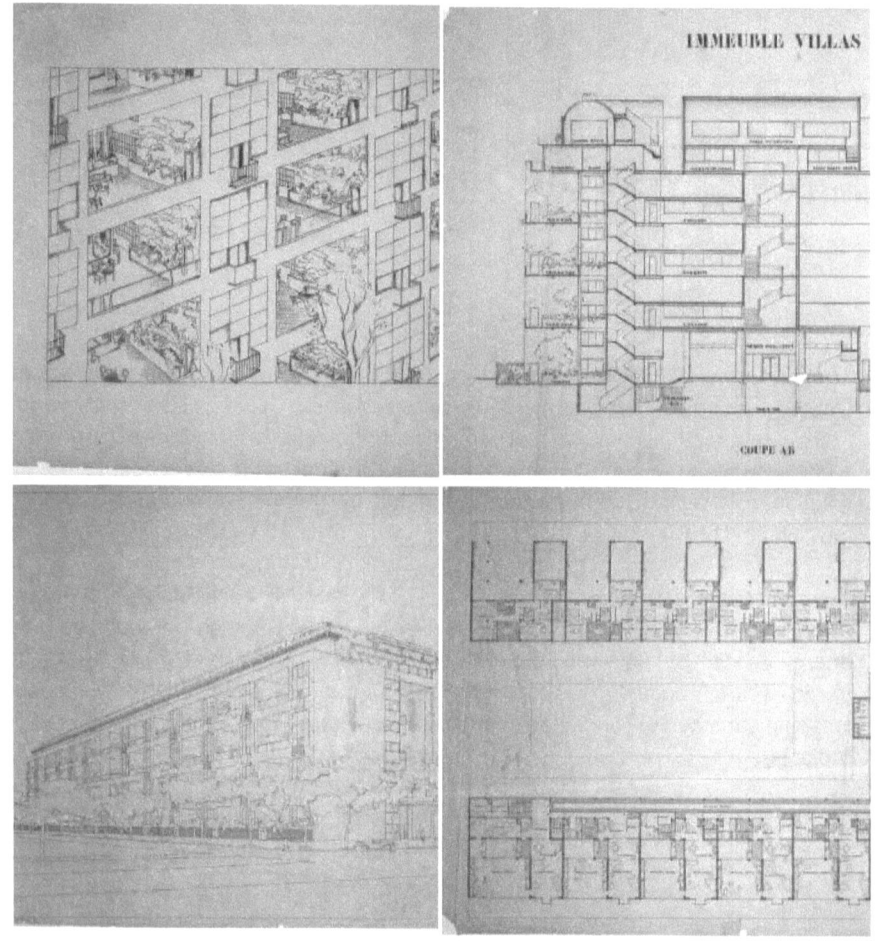

Immeubles-Villas designed by Le Corbusier. Source Foundation Le Corbusier / ADAGP

James Wines SITE Highrise of Homes (1981)

Multi storey building with vertical community of private houses, clustered into district village-like on each floor

High-rise of homes. Source: SITE / James Wines

Ken Yeang's Bioclimatic Skyscraper (Menara Mesiniaga, built 1992)

Fifteen-storey office building constructed in 1992 in Malaysia. This innovative structure integrates bioclimatic design principles, utilising natural air flow for ventilation and featuring sky gardens on various floors.

Mesiniaga Tower viewed from the Subang–Kelana Jaya Link. Source: *Wikipedia* / Cmglee

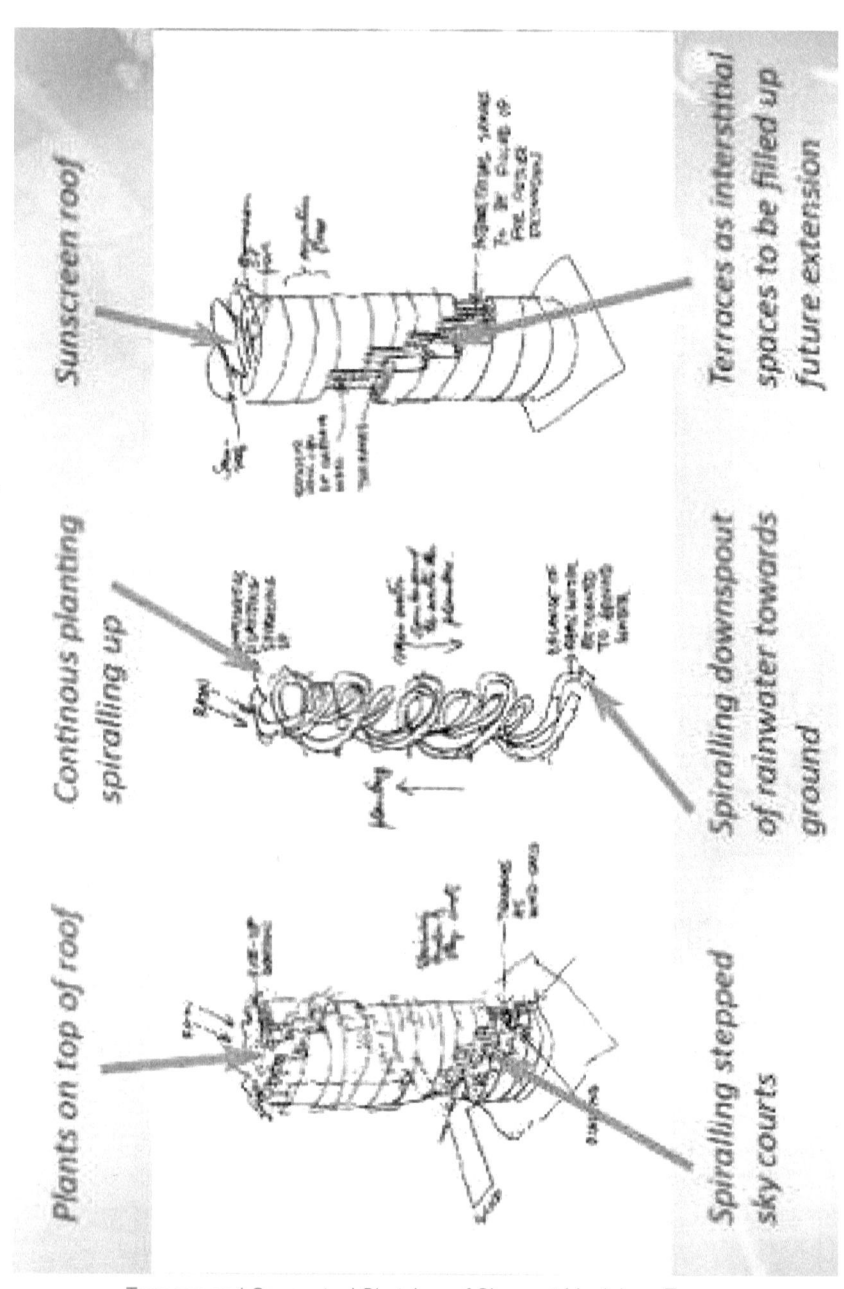

Terraces and Conceptual Sketches of Skycourt Mesiniaga Tower.
Source: Ken Yeang / Suci Farahdilla, 2021

Vincent Callebaut's Dragonfly (2009)

Metabolic farm for urban agriculture, designed to resemble a dragonfly with two large glass wings, this structure combines residential, office, and agricultural spaces in a single building.

View of the Dragonfly building from the Empire State Building.
Source: Vincent Callebaut Architectures

Interior view of the greenhouse between the two dragonfly wings.
Source: Vincent Callebaut Architectures

MVRDV Bagneux (2017)

A neighbourhood that integrates urban farming solutions to meet the demand for locally produced food.

Key features include a green tower for urban farming, sunny gardens, and outdoor spaces for each apartment, promoting self-sufficiency and enhancing living quality.

A detailed architectural rendering of Bagneux highlights a sustainable urban environment featuring innovative roof gardens and urban farming / Source: MVRDV

Modern buildings with greenery on rooftops and terraces, integrating plants and vegetables across multiple levels. / Source: MVRDV

Superfarm Project by Studio NAB (2019)

Six-story floating urban farming tower focuses on high-yielding foods and includes a variety of production techniques such as aquaponics, insect farming, and seaweed culture.

The project aims to bring agriculture closer to urban residents and promote sustainable urban living.

General view of the floating urban farming tower. Source: Studio nab / Nicolas Abdelkader

Internal view of the floating urban farming tower with hydroponics and aquaponics structures. Source: Studio nab / Nicolas Abdelkader

Ilimelgo Vertical Urban Farm (Romainville, France, built 2021)

Seven-story structure emphasises educational purposes and community engagement. The farm incorporates natural light for plant growth, avoiding the use of LEDs. It houses a restaurant, workshops, and pedagogical gardens on the ground floor

Street view of Vertical Urban Farm in Romainville. Source: Ilimelgo / Sandrine Marc

Internal view of Vertical Urban Farm in Romainville. Source: Ilimelgo / Sandrine Marc

van Bergen Kolpa Architecten Vertical Farm (Hengshui, built 2023)

Four layer research and demonstration centre: the first two for automated vertical crop production under LED lighting, laboratories, and processing areas; the third floor for horticultural suppliers; and the rooftop Sky Farm for horizontal vegetable cultivation.

Street view of Vertical Farm in Hengshui. Source: van Bergen Kolpa Architecten

Internal view of Vertical Farm in Hengshui. Source: van Bergen Kolpa Architecten

Architects and plant scientists have been working together for a long time to **create new ways of bringing farms and nature into cities**.

They've come up with new ideas for buildings that aren't just places to live or work, but also places to grow food and enjoy nature.

> Even though some of these ideas were thought up many years ago and some projects were left on paper and never built, they've helped **create the vertical farms and city gardens we see today.**

They show us that **we can have cities that are good for both people and plants**.

How Vertical Farming Has Changed

Initial attempts at vertical farming were **small-scale**, involving simple setups with plants grown on shelves under daily and artificial lights.

These early experiments were crucial in proving that **we could successfully grow plants in controlled indoor environments**.

Two panels from the first *Suske en Wiske* (English: Spike and Suzy) comic strip *Op het eiland Amoras* by Willy Vandersteen depict a primitive vertical farm. The Belgian comic strip was published in *De Nieuwe Standaard* from 19, 1945 December to 13 May 1946. Its caption reads: "**Every patch of cultivated land is storeyed and artificially watered and lit, yielding a double harvest**." With copyright permission from Standaard Uitgeverij. Source: Van Gerrewey et al., 2022

Technological advancements have been the real game-changer for vertical farming

LED (Light Emitting Diode) technology, which can be tuned to specific wavelengths for optimal plant growth, has made indoor farming more energy-efficient.

Automated irrigation systems have ensured that plants receive the precise amount of water and nutrients needed.

Advanced computer monitoring has enabled farmers to track and adjust growing conditions in real-time.

These technologies have improved the efficiency and scalability of vertical farms, reducing resource use while increasing crop yields.

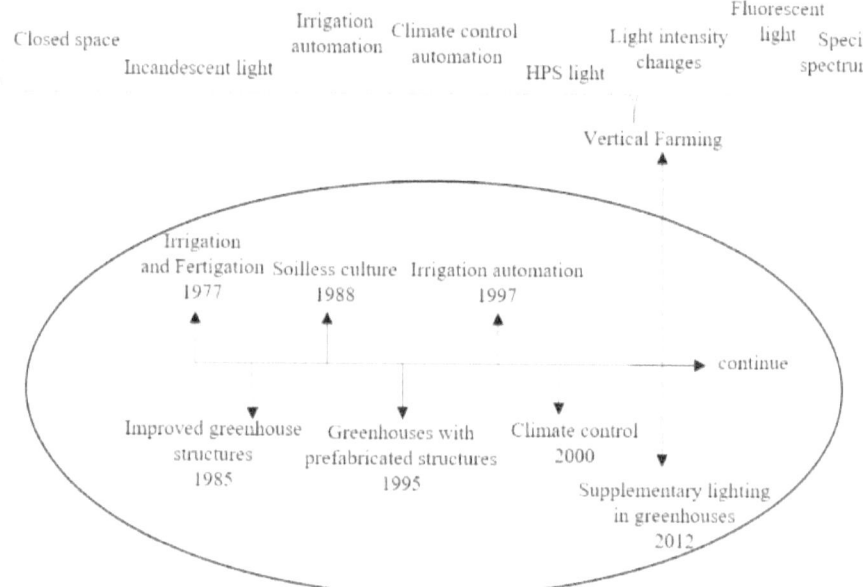

Innovations in intensive agriculture and major developments in vertical farming systems in line with the sustainability of agriculture. Source: Nájera et al., 2023

As the concept gained popularity, businesses saw the potential. Companies began converting warehouses and old industrial buildings into **large-scale vertical farms**. The methods used in vertical farming have also expanded.

Vegetable factory production line of Fujitsu AkiSai; (b) Swiss Rogue/s vertical farming automation system; (c) AeroFarms vertical farm. Source: Min et al., 2023

Global Relevance and Potential

Vertical farming is gaining significance worldwide due to various factors that make it a promising solution for contemporary agricultural challenges.

One of the primary drivers of vertical farming's relevance is the **growing population**.

> By 2050, the world's population is projected to reach nearly 10 billion people, necessitating a **substantial increase in food production**.

Vertical farms, with their ability to maximise space efficiency by stacking layers of crops, can help meet this rising demand. They can be established within cities, transforming urban environments into hubs of agricultural production.

Urban expansion has resulted in the loss of valuable farmland. As cities grow, they encroach upon rural areas traditionally used for agriculture. Vertical farming mitigates this issue by enabling food production within urban settings, thus preserving existing farmland and reducing the need for further land conversion. This urban-centric approach not only conserves land but also brings food production closer to consumers.

Climate change poses significant challenges to traditional farming, with unpredictable weather patterns and extreme events disrupting crop yields. Vertical farming offers a controlled environment where temperature, humidity, and light can be regulated, ensuring stable and predictable crop production regardless of external weather conditions. This resilience makes vertical farming an attractive option in the face of a changing climate.

Access to fresh food is another critical benefit of vertical farming. Urban residents often rely on food transported over long distances, leading to decreased freshness and nutritional value. Vertical farms located within cities can provide local populations with fresher produce, reducing transportation costs and the associated environmental impact. This proximity also supports local economies and enhances food security.

Scientific Report

How Vertical Farms and Greenhouses Can Influence the Sustainability and Footprint of Urban Microclimate with Local Food Production?

Country: Greece
Publication Date: 8 August 2022
Main Focus: The study systematically reviews how controlled-environment agriculture (CEA) methods, particularly vertical farms (VFs) and greenhouses (GHs), can enhance sustainability and reduce the environmental footprint in urban microclimates through local food production.

Key Findings:

- VFs and GHs significantly contribute to reducing CO_2 emissions by minimising food transportation and post-harvest processes.
- VFs have the potential to revolutionise urban food systems but require improvements in energy efficiency, especially concerning artificial lighting.
- Large-scale VF implementation with renewable energy sources could optimise urban food and energy systems, promoting sustainable urban living

Reference: Vatistas, C.; Avgoustaki, D.D.; Bartzanas, T. A Systematic Literature Review on Controlled- Environment Agriculture: How Vertical Farms and Greenhouses Can Influence the Sustainability and Footprint of Urban Microclimate with Local Food Production. *Atmosphere* 2022, *13*, 1258. DOI: 10.3390/atmos13081258

Water scarcity is a pressing issue in many regions, and agriculture is a major consumer of water resources. Vertical farming systems, such as hydroponics and aeroponics, use significantly less water than traditional soil-based agriculture. By recirculating water and delivering it directly to plant roots, these systems achieve high water efficiency, making vertical farming a sustainable option in water-scarce areas.

Vertical farming also enables year-round food production, independent of **seasonal variations**. This is particularly beneficial in regions with short growing seasons or harsh winters, where traditional farming is limited to certain times of the year. Continuous production ensures a steady supply of fresh produce, enhancing food availability throughout the year.

The advent of vertical farming creates new **job opportunities** in urban areas. As the industry grows, it demands a workforce skilled in both agriculture and technology. This intersection of fields provides a promising career path for young people, offering roles in farm management, system maintenance, and agricultural research.

Moreover, vertical farming can contribute to **reducing pollution**. Traditional agriculture involves extensive transportation of food products from rural areas to urban markets, contributing to greenhouse gas emissions and air pollution. By localising food production, vertical farming minimises the need for long-distance transport, thus reducing the carbon footprint associated with food distribution.

Food safety is another advantage of vertical farming. The controlled environments within vertical farms are less susceptible to pests and diseases, reducing the need for chemical pesticides and herbicides. This leads to cleaner, safer food with lower risks of contamination. Additionally, vertical farms can implement stringent hygiene practices, further enhancing food safety standards.

Educational opportunities abound with vertical farming. Schools can establish small-scale vertical farms as teaching tools, integrating them into science and technology curricula. These projects can inspire students to explore careers in agriculture, biotechnology, and environmental science, cultivating a new generation of innovators and problem-solvers.

> However, vertical farming is not without its challenges. The **initial cost** of setting up a vertical farm can be high, and the systems require a **considerable amount of electricity** to power lights, climate control, and automated systems. Additionally, **not all crops** are suitable for vertical farming, as some require extensive space or specific growing conditions that are difficult to replicate indoors.

Even though vertical farming has some problems, its **future looks good**. New technology keeps making it better and easier to do. More people are learning about it and how it works. As it improves, vertical farming could help feed more people around the world in a way that's good for the planet.

2. The Science Behind Vertical Farming

This chapter, *The Science Behind Vertical Farming,* explores the fundamental principles and methods that underpin this innovative agricultural approach.

Plant Biology Basics explain the groundwork for understanding **how plants grow and thrive in controlled environments**. This section covers essential topics such as photosynthesis, nutrient uptake, and provides readers with the necessary biological context for vertical farming practices.

To understand the basics of vertical farming, a *Comparing Indoor and Outdoor Vertical Farming* shows the **unique challenges and advantages** of each approach. This comparative analysis helps to understand the rationale behind choosing one method over the other, depending on factors such as climate, available resources, and specific crop requirements.

There is also a dive into *Soilless and Substrate-based Growing Methods*, widely used in vertical farming, such as **hydroponics, aeroponics, and aquaponics** and exploration of the main principles behind these techniques.

Plant Biology Basics

The science of vertical farming is rooted in understanding **plant biology**.

Plants are **complex organisms** that have evolved to thrive in various environments. In vertical farming, we apply this knowledge to create optimal growing conditions.

Relevant parameters related to plant morphological and physiological in Vertical Farming systems. Source: Najera et al., 2023

Parameter	Description
Photosynthetic parameters	Photosynthetic parameters are among the most commonly used physiological metrics in crop studies. These parameters are non-destructive and help accurately understand plant ecophysiology.
Leaf Area (LA)	Leaf area is a crucial parameter that determines the amount of photoassimilates produced, impacting plant growth, development, and productivity. Generally, leaf area and biomass accumulation in crops can be influenced by factors such as light quality and radiation intensity, plant nutrition, substrate type, container design / volume.
Plant productivity	Plant productivity is often measured by fresh weight, which is of commercial interest due to its relevance in the market. However, researchers focus on dry weight for every plant organ (stem, leaf, root, or fruit), as biomass accumulation results from photosynthetic activity and CO2 concentration.
Total Dry Matter (TDM)	Total dry matter represents the net gain in dry matter and is considered one of the best indicators of plant quality. Plants with high TDM content exhibit significant growth potential and high field yield.

At the core of plant biology are several key processes.

Photosynthesis is fundamental - plants use light energy to convert carbon dioxide and water into glucose and oxygen. The core components of the photosynthesis process are chlorophyll, light, water, carbon dioxide, oxygen, glucose, and chloroplasts.

In vertical farms, artificial lighting is calibrated to provide the specific wavelengths that best support photosynthesis.

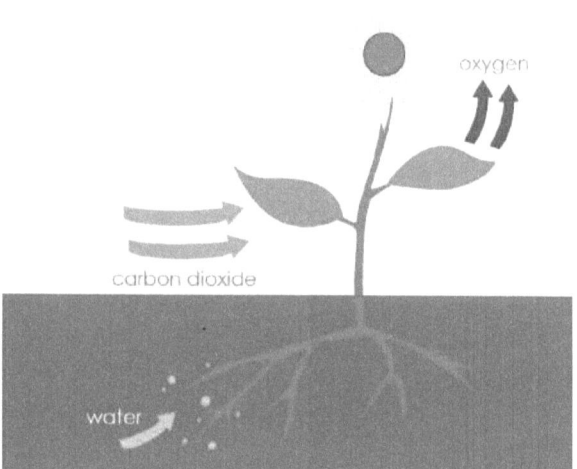

Photosynthesis. Source: *Wikimedia Commons*

Plant growth and development depend on more than just light. Plants need 16 essential elements for growth and development [Xu et al., 2021].

Carbon, hydrogen, and oxygen are obtained from air and water, while the other essential elements are absorbed from the soil through the roots.

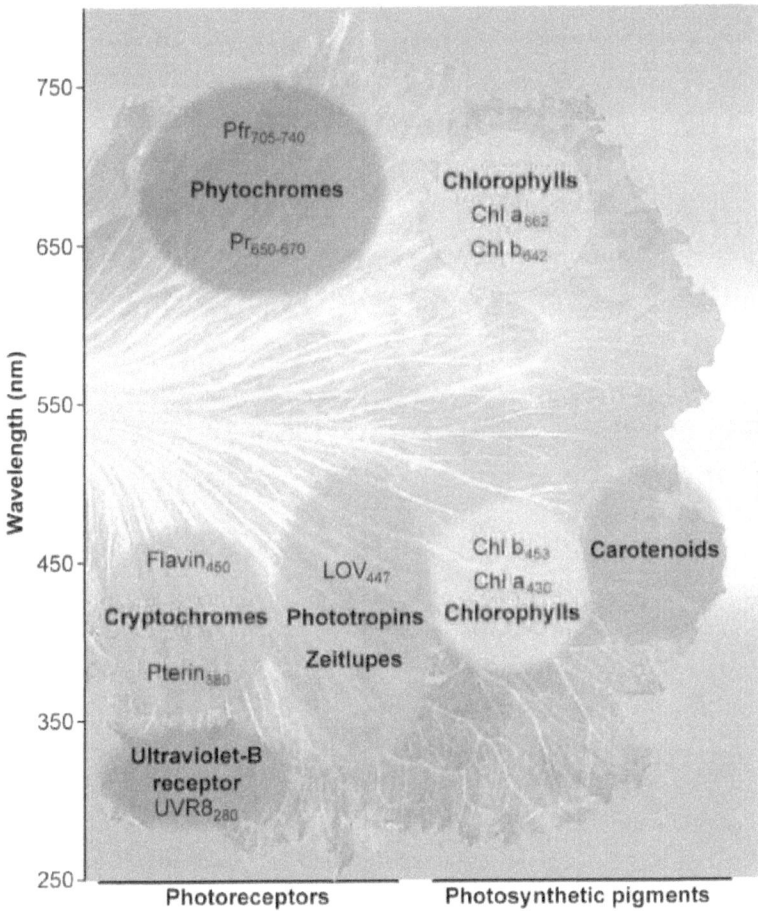

Details of photosynthesis and light perception in plants. It details how different photosensitive molecules absorb specific wavelengths of light, which is crucial for processes like energy capture in photosynthesis (via photosynthetic pigments like chlorophylls) and regulation of plant growth and development (via photoreceptors like phytochromes and cryptochromes). Source: Wong *et al*, 2020

The main nutrients needed in large quantities, known as **macronutrients**, include nitrogen, phosphorus, potassium, calcium, magnesium, and sulphur.

To ensure they get these vital nutrients, plants have developed various mechanisms to efficiently absorb and use these ions. The dynamics of nutrient uptake rates refer to how plants absorb macronutrients (such as nitrogen, phosphorus, and potassium) and secondary nutrients (like calcium, magnesium, and sulfur) over time, **affecting their growth and development**.

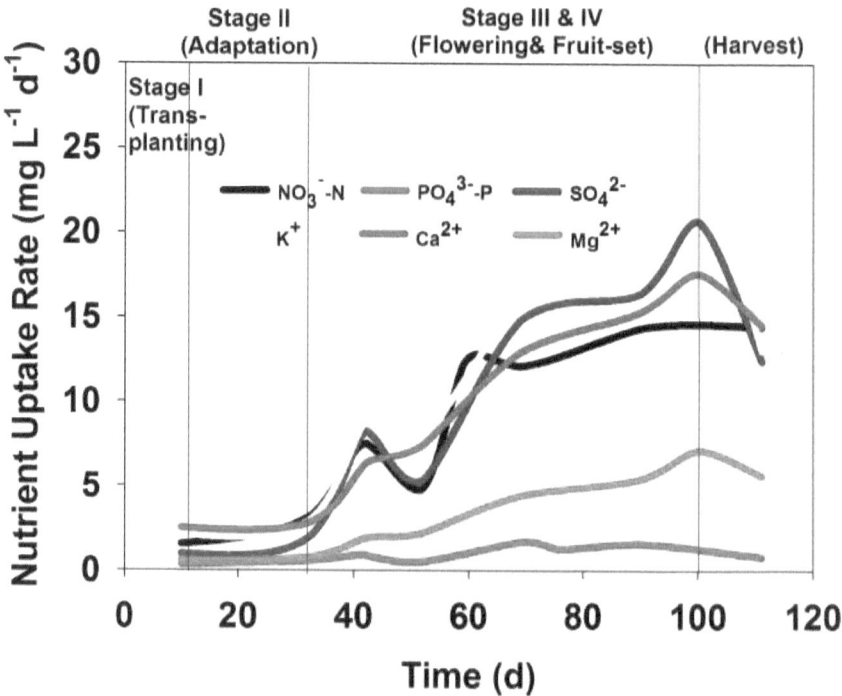

Dynamics of nutrition uptake rates of macro nutrients and secondary nutrients during tomato growth in the closed hydroponic system. Source: Lee et al., 2017

Water and nutrient uptake are important for plant growth and health. In traditional soil systems, roots must explore varying soil conditions to access nutrients and water.

In hydroponic and aeroponic systems, nutrients are **delivered directly in a solution**, allowing for precise control and optimal absorption.

This leads to faster growth, higher yields, and reduced risk of soil-borne diseases, as plants receive consistent and efficient nutrient delivery without the variability and limitations of soil.

Water and nutrient uptake primarily occurs **through the roots**.

Therefore, it is **important to minimise any restrictions** on root growth in soilless agriculture to ensure efficient absorption and overall plant well-being.

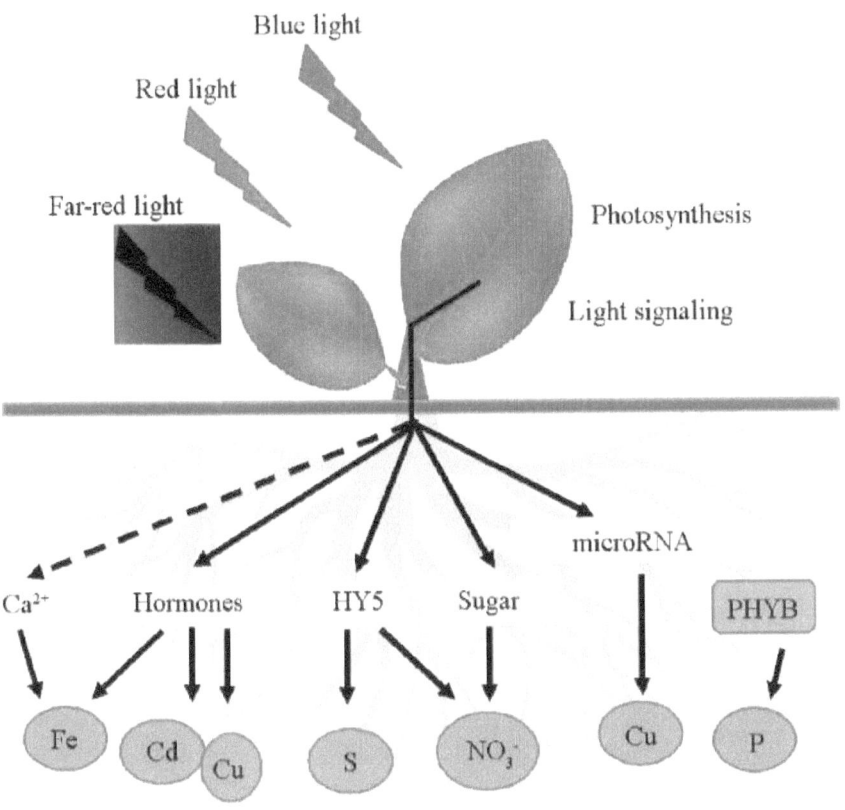

Model showing multiple light signalling pathways regulating nutrients uptake.
Source: Xu et al., 2021

Plant responses to root restriction in soilless agriculture.
Source: Balliu et al., 2021

Plant Response	Crop	Soilless Production System	Additional Information
Reduced dry matter of roots	Chilli pepper	Polyvinyl-chloride (PVC) columns, filled with a mixture of coconut coir dust and empty fruit bunch compost (70:30, v:v)	9570 mL (control) vs. 2392 mL (root-restricted) columns

Plant Response	Crop	Soilless Production System	Additional Information
	Pepper	Plastic pots (three seeds per pot) containing Fafard 2B mix (Sun Gro Horticulture, Agawam, MA)	500 mL (control) vs. 60 mL (restricted) containers
	Cucumber	Floating system (F.S.)	Control vs. 40 mL (restricted) vessels
Adventitious Root formation	Cucumber	Floating system (F.S.)	Control vs. 40 mL (restricted) vessels
	Tomato	Flow-through hydroponic culture system (FTS)	1500 mL (control) vs. 25 mL (restricted) containers
Dense mat of roots	Cucumber	Floating system (F.S.)	Control vs. 40 mL (restricted) vessels
	Tomato	Flow-through hydroponic culture system (FTS)	1500 mL (control) vs. 25 mL (restricted) containers
	Sweet potato	A mixed system of solid media and nutrient solution	4.5 L, 3.0 L, and 1.6 L pots
Yield reduction	Tomato	Different alternatives of solid growing media (perlite, pumice, volcanic ash, perlite + peat, pumice + peat, volcanic ash + peat)	8 L and 4 L pots
	Processing tomato	Solid growing media (Metro-Mix 350, Sun Gro Horticulture)	26 L, 16, 6, and 1 L pots

Plant Response	Crop	Soilless Production System	Additional Information
Non-significant yield reduction	Pepper	Growth media (Fafard 2B mix; Sun Gro Horticulture, and Turface clay) mixed in a 3:1 ratio	1500 mL, 500 mL, and 250 mL plastic pots
	Tomato	Coconut fibre substrate	10, 7.5 and 5 L pots
Increased harvest index	Pepper	Growth media (Fafard 2B mix; Sun Gro Horticulture, and Turface clay) mixed in a 3:1 ratio	1500 mL, 500 mL, and 250 mL plastic pots
	Chilli pepper	Polyvinyl-chloride (PVC) columns, filled with a mixture of coconut coir dust and empty fruit bunch compost (70:30, *v:v*)	9570 mL (control) vs. 2392 mL (root-restricted) columns

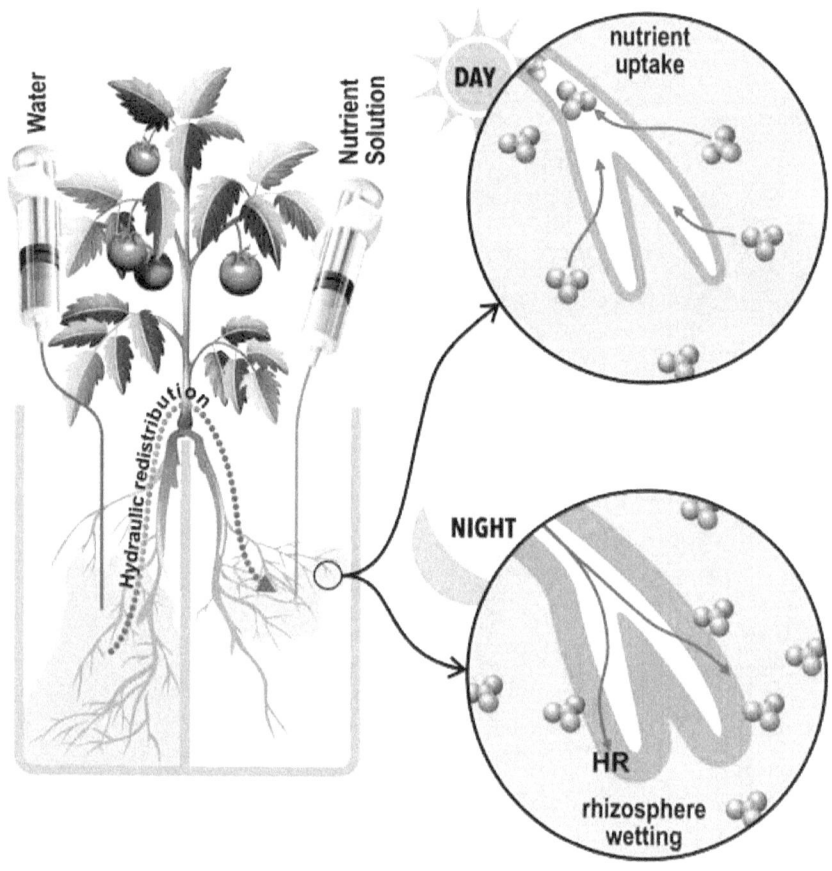

Root uptake under mismatched distributions of water and nutrients in the root zone.
Source: Yan et al., 2020

Scientific Report

Root Uptake Under Mismatched Distributions of Water and Nutrients in the Root Zone
Country: United States 🇺🇸
Publication Date: 17 December 2020
Main Focus: The study investigates how plant roots adapt to environments where water and nutrients are unevenly distributed in the soil.

Key Findings:
- Plants can thrive even when water and nutrients are not evenly distributed, utilising mechanisms like localised root proliferation in nutrient-rich, dry soil patches.
- Roots can transfer water from wetter areas to drier, nutrient-rich areas, enhancing nutrient uptake and preventing root stress.
- The presence of essential nutrients influences root growth patterns and the development of root hairs, which aids in nutrient uptake.
- Separating the application of water and nutrients can improve nutrient use efficiency and reduce water pollution in intensive agricultural systems.

Reference: Yan, J., Bogie, N. A., & Ghezzehei, T. A. Root uptake under mismatched distributions of water and nutrients in the root zone. *Biogeosciences* **2020**, 17*(24), 6377–6392. DOI: 10.5194/bg-17-6377-2020

Environmental and cultivation factors affect morphology, architecture and performance of root system in soilless grown plants

- Root restriction
- Nutrient solution
- Water supply
- Temperature, pH
- Oxygenation
- Beneficiary microorganisms
- Root exudates, allelopathy
- VPD
- Lighting, CO_2

Environmental and cultivation factors affect the morphology, architecture, and performance of root systems in soilless grown plants. Source: Balliu *et al.*, 2021

Plant responses to irrigation frequency in soilless agriculture. Source: Balliu *et al.*, 2021

Plant Response	Crop	Soilless Culture System	Additional Information
Increased irrigation frequency increases plant yield	Chrysanthemum	Seedling tray contained coconut peat	Irrigation frequencies of 4, 6, and 8 times/day
	Tomato	40-L (15 cm × 18 cm × 120 cm) bags containing expanded perlite	Irrigation applied when the plants had consumed 0.4-, 0.8-, or 1.2-L of water

Vertical root-density distribution mimics container moisture content. Denser at the lower part of the container.	Tomato	Wood fibre substrate	
	Chili pepper	31 × 15 × 60 cm container filled with sandy-loamy soil	1-, 3-, and 5-day irrigation intervals

Plant responses to nutrient solution. in soilless agriculture.
ource: Balliu et al., 2021

Plant Response	Crop	Soilless Culture System	Additional Information
No increase in plant yield by increasing N fertilization rates	Lettuce	Washed sand; 2.5 L (no confinement, the control); 1.0 L (moderate) and 0.4 L (severe root restriction)	Total nitrogen concentrations in mM L1, 5.55, 8.05, 10.55, 13.05 and 15.55.
	Spinach	Styrofoam trays floated into 80 cm × 44 cm × 19 cm (52 L) plastic basins	"Full dose" nutrient solution (mg L−1: N 150, P 50, K 150, Ca 150, Mg 50, Fe 5.0, Mn 0.50, Zn 0.05, B 0.50, Cu 0.03, Mo 0.02), "half dose" (with macro elements reduced by 50%)
	Baby leaf lettuce	Styrofoam trays floated into 135 cm × 125 cm × 20 cm a flotation bed	Nutrient solutions with 12 and 4 mM L−1 N
Primary root growth inhibition, increase in lateral roots and root hairs	Various crops	Various production system	Limited P supply
Increase in vertical, deep roots	Various crops	Various production systems	Limited N supply
Increased root dry weight, specific root length, root tissue density, and root length density due to	Tomato, Zinnia	450 mL plastic pots containing either Metromix 360 (MM360) or Ball	24-h, 48-h, and 96-h irrigation intervals

Plant Response	Crop	Soilless Culture System	Additional Information
increased irrigation intervals		Professional Growing Mix (BPGM)	
	Chili pepper	31 cm × 15 cm × 60 cm container filled with sandy-loamy soil	1-, 3- and 5-day irrigation intervals

The plant's vascular system, composed of **xylem** and **phloem**, is crucial for **internal transport**. Xylem moves water and dissolved minerals upward from the roots, while phloem distributes the products of photosynthesis throughout the plant.

One more component of the plant's transport system is the **stomata**. The plant has the ability to **regulate water loss and gas exchange** through it. Stomata play a crucial role in the allocation of water, nutrients, and signals through the process of transpiration.

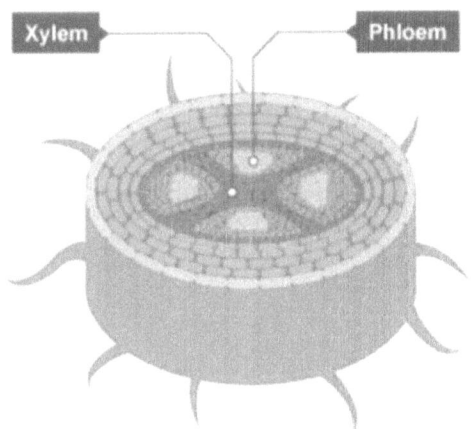

Xylem and phloem.
Source: *BBC* / GCSE Biology

When stomata are open, they allow the **free flow of hydraulic content**, including water, nutrients, and signals, from the root xylem into the stem xylem vessels and throughout the plant. This also facilitates the **osmotic flow** from xylem to phloem vessels. Conversely, when stomata are closed, transpiration is suppressed, reducing the flow of hydraulic content between the root and stem xylem vessels and slowing the osmotic transfer to the phloem vessels [Twalla *et al.*, 2022].

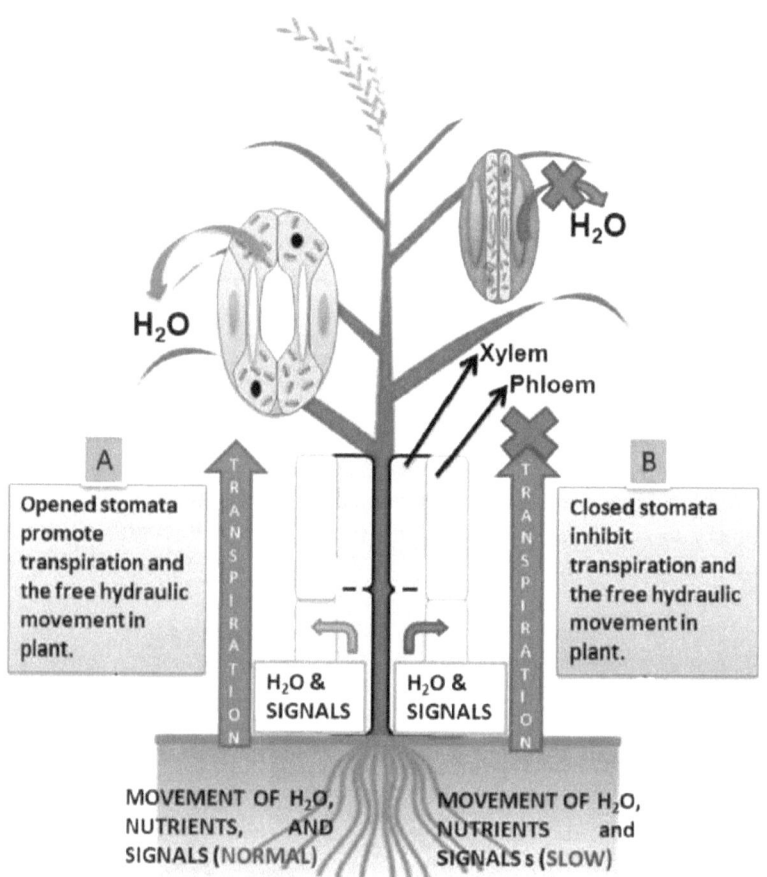

A schematic representation of how stomata regulate the movement of water, nutrients, and signals in plants through transpiration. Source: Twalla et al., 2022

The safety-efficiency trade-off in plants refers to the **balance between maximising the efficiency of resource use** (such as water and nutrients) **and minimising the risk of damage or stress**.

In the context of stomatal function, efficiency involves **maximising the uptake of carbon dioxide** for photosynthesis and the effective transport of water and nutrients, while safety involves **minimising water loss and avoiding damage** under stressful conditions, such as drought.

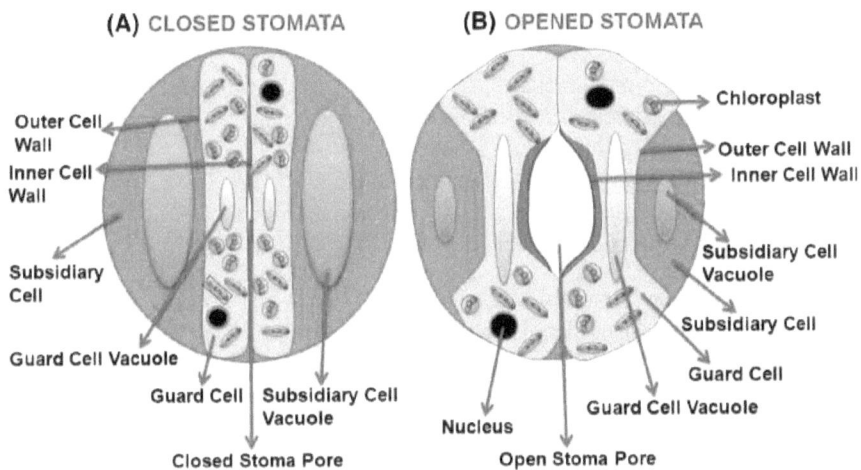

The diagram shows the basic structure of open and closed stomatal complexes in grass cereals. It highlights parts like subsidiary cells, vacuoles, chloroplasts, cell walls, guard cells, the stomatal pore, and nuclei. In diagram A, the stomatal pore is closed, with expanded subsidiary cells and a reduced guard cell size due to low water. In diagram B, the stomatal pore is open, with reduced subsidiary cells and expanded guard cells due to higher water levels. Source: Twalla et al., 2022

Scientific Report

A Stomatal Safety-Efficiency Balance Constraints Responses to Leaf Dehydration
Country: United States
Publication Date: 26 July 2019
Main focus: The study investigates the trade-off between stomatal conductance and safety in plants, particularly how these traits balance the needs for gas exchange and drought resistance.

Key findings:

- The research identifies a trade-off between high maximum stomatal conductance (gmax) and the sensitivity of stomatal closure during dehydration ($\Psi gs50$). Plants with high gmax, which are efficient at gas exchange under well-watered conditions, are more sensitive to drought, showing higher rates of stomatal closure as leaf water potential decreases.
- The study supports this trade-off through experiments on various Californian woody species and analyses of global plant species data, linking stomatal traits to both physiological and structural factors like stomatal size, density, and osmotic concentration.

- This trade-off influences water use and drought tolerance across species, impacting their ecological strategies and potential responses to climate change.
- The findings suggest that this safety-efficiency trade-off is a general feature across different plant species and may play a critical role in shaping plant distribution and function in various environments.

Reference: Henry, C., John, G.P., Pan, R. et al. A stomatal safety-efficiency trade-off constrains responses to leaf dehydration. *Nat Commun* **2019**, 10, 3398. DOI: 10.1038/s41467-019-11006-1

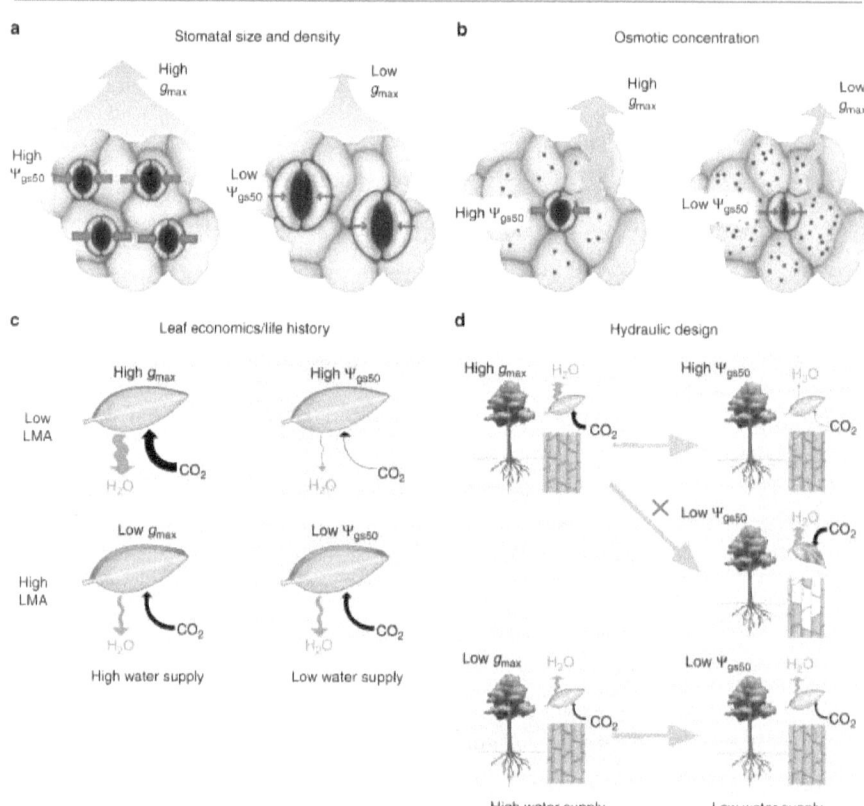

The hypothesised reasons for a stomatal safety-efficiency trade-off suggest that smaller, denser stomata improve conductance but increase drought sensitivity. Weaker osmotic concentrations also boost conductance. Species with lower leaf mass per area have higher photosynthesis but are more sensitive to water supply changes, while high conductance species reduce photosynthesis to prevent damage during water shortages.
Source: Henry et al., 2019

Plant hormones play a vital role in growth and development.

Auxins, cytokinins, gibberellins, and other hormones regulate processes like **stem elongation, leaf expansion, and flowering**.

In vertical farming, **manipulating environmental factors** can influence hormone production and distribution, allowing for more control over plant growth patterns.

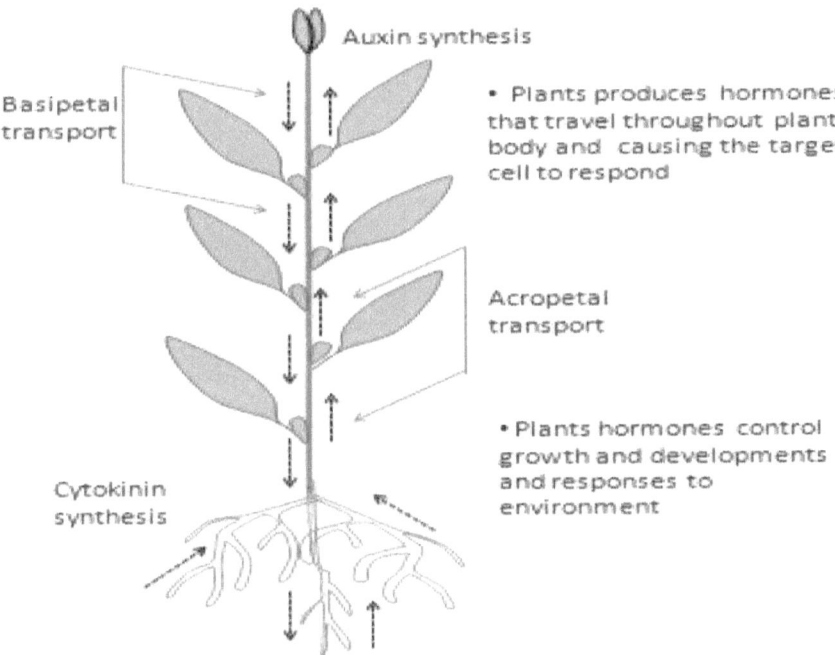

Role of hormones in plant growth and development. Source: Khan *et al.*, 2020

Plants also **respond to environmental stimuli** through tropisms.
Phototropism (response to light) and **gravitropism** (response to gravity) are particularly relevant in vertical setups.

The orientation of growth as a function of gravity and light, respectively, are two major processes implicated in the regulation of **plant shape** [Coutand et al., 2019].

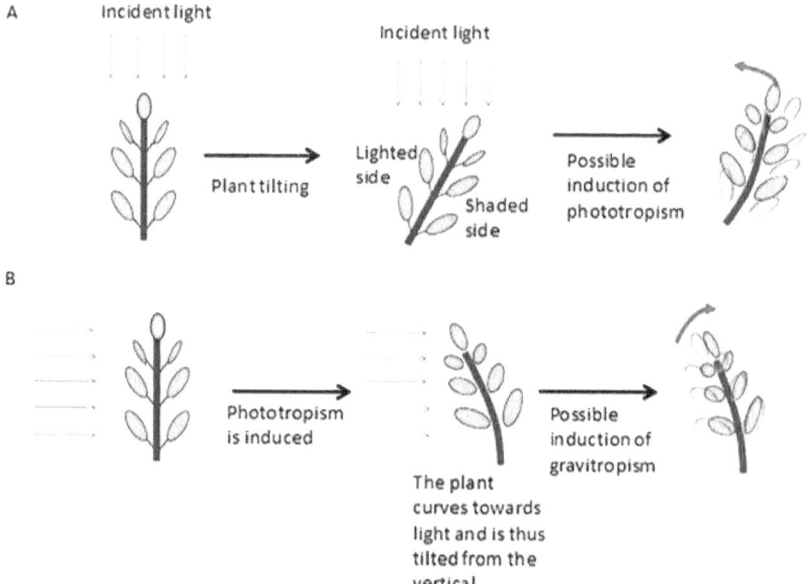

Diagram explaining the interactions between gravitropism and phototropism in classical experiments dedicated to the study of gravitropism and phototropism.
A: In classical experiments, a plant is tilted from the vertical in order to study gravitropism. In this case, since the light is coming from the top, one side of the stem is shaded, which can induce a phototropic reaction.
B: In classical experiments, a plant is lit unilaterally in order to study phototropism. Since the plant bends towards the unilateral light, the stem is curved and thus tilted, which can induce a possible counteractive movement due to gravitropism.
Source: Coutand et al., 2019

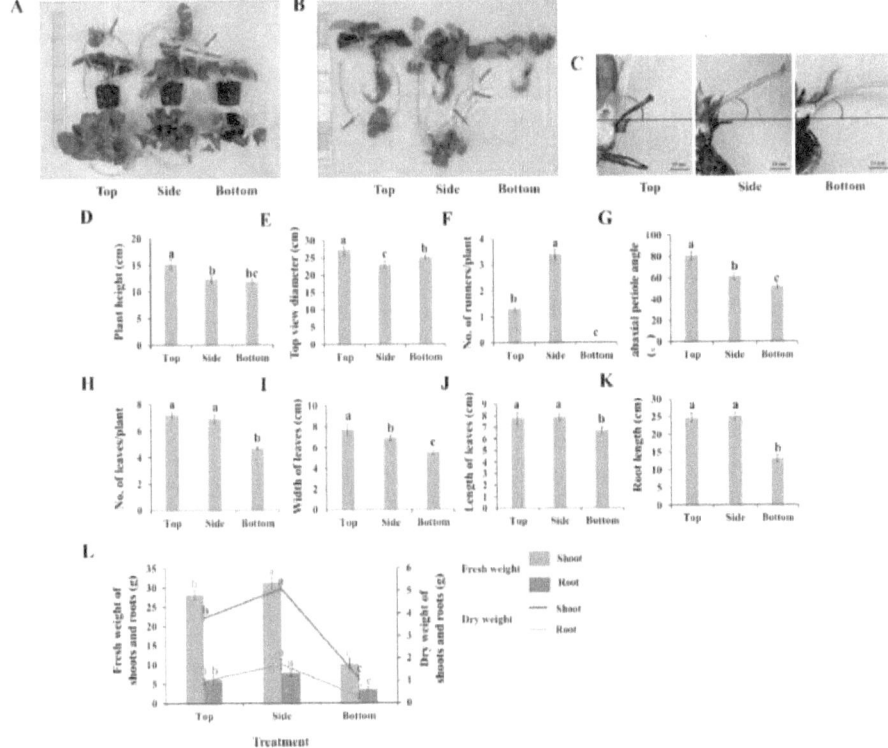

Changes in the phenotype and plant traits of strawberry 'Sulhyang' plants as affected by the lighting direction after 45 days of cultivation.
The plant morphology (**A**), root state and length (**B,K**), abaxial petiole angle of the outermost leaves (**C,G**), plant height (**D**), top view diameter of plants (**E**), number of runners per plant (**F**), number of leaves per plant (**H**), length and width of the largest fully-expanded leaves (**I,J**), fresh and dry weights of plant shoots and roots (**L**).
Vertical bars indicate the means ± standard error ($n = 6$).
Different lowercase letters indicate significant separations within treatments by Duncan's multiple range test at $p \leq 0.05$. Bars indicate 10 mm.
Red arrows indicate runners per plant.
Source: Yang et al., 2022

Scientific Report

Side Lighting Enhances Morphophysiology and Runner Formation by Upregulating Photosynthesis in Strawberry Grown in Controlled Environment
Country: South Korea
Publication Date: 23 December 2021
Main focus: This study investigates how different lighting directions affect the growth, development, and runner formation of strawberries grown in controlled environments.

Key findings:
- Side lighting significantly enhanced plant morphophysiology and runner formation by upregulating photosynthesis.
- Side lighting resulted in the highest photosynthetic rate (14.4 µmol CO_2 $m^{-2}·s^{-1}$) and stomatal conductance (0.67 mol H_2O $m^{-2}·s^{-1}$) compared to top and bottom lighting.
- The study demonstrated that side lighting can promote commercial benefits, such as reduced economic costs and easier controllability, without harming the plants.

Reference: Yang, J.; Song, J.; Jeong, B.R. Side Lighting Enhances Morphophysiology and Runner Formation by Upregulating Photosynthesis in Strawberry Grown in Controlled Environment. *Agronomy* **2022**, *12*, 24. DOI: 10.3390/agronomy12010024

Lastly, understanding **plant stress responses** is important for decision-making.

> **Biotic stress** is caused by living organisms, such as pathogens, insects, or weeds, which can lead to diseases or pest infestations. **Abiotic stress**, on the other hand, is caused by non-living environmental factors like drought, extreme temperatures, or soil salinity.

Increasing the number of different co-occurring multifactorial stress factors **causes a severe decline in plant growth and survival**, as well as in the microbiome biodiversity that plants depend upon [Zandalinas *et al.*, 2021]

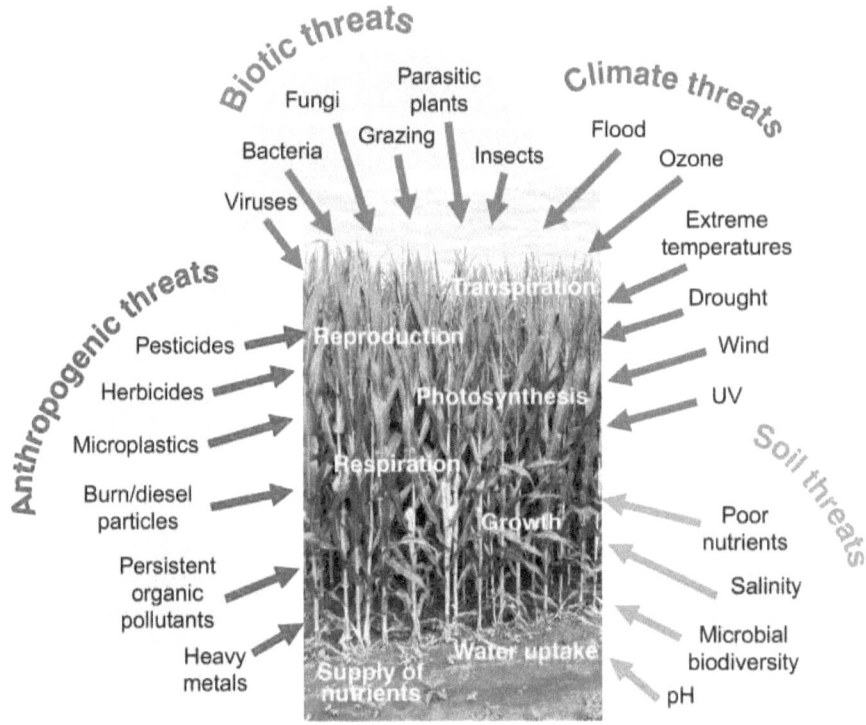

Definition of a multifactorial stress combination. Source: Zandalinas et al., 2021

Vertical farming systems are designed to **minimise these stressors**, creating an environment where plants can direct more energy towards growth and production rather than survival mechanisms.

By applying principles of plant biology, **vertical farming creates optimised environments** that support plant growth.

This scientific approach allows precise control over factors like light, water, and nutrients, leading to higher yields, faster growth cycles, and more efficient resource use compared to traditional agriculture. Vertical farming represents a blend of biology and technology, paving the way for sustainable and productive urban agriculture.

Comparing Indoor and Outdoor Vertical Farming

There are various classifications of growing settings and their types that are relevant to the category of vertical farming.

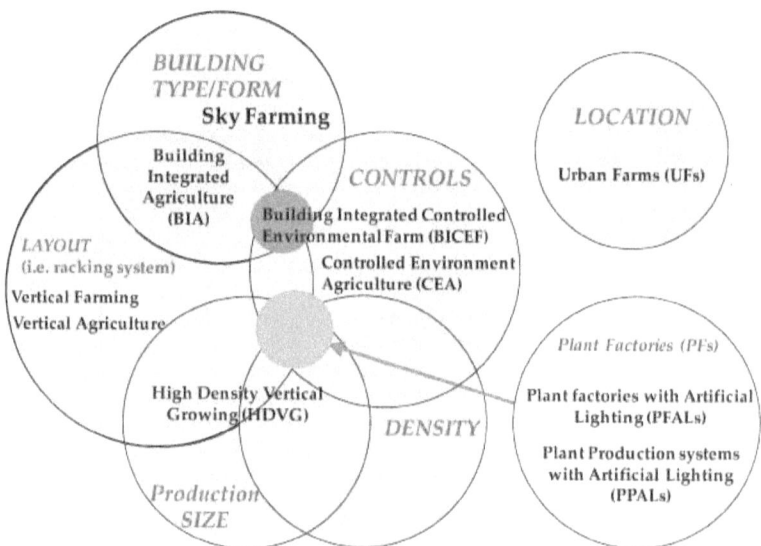

Types of vertical farms and influencing factors. Source: Kabir et al., 2023

Two main categories of vertical farming are **indoor vertical farms and outdoor vertical gardens**, each with its unique features and applications.

Both indoor and outdoor vertical farming aim to **maximise space efficiency** in urban areas, but they differ significantly in their **resource utilisation, environmental impact,** and **integration with existing urban infrastructure**.

> The main difference between indoor and outdoor vertical farming systems is **how much control** they have over the environment and their use of natural resources.

Indoor vertical farms are enclosed systems that offer complete control over the growing environment, providing an innovative solution to urban agriculture challenges. These facilities can be housed in **various structures**, including purpose-built warehouses, repurposed industrial buildings, and even skyscrapers in dense urban areas.

These systems offer **complete control over climate, lighting, and nutrient delivery**, allowing for year-round production regardless of external weather conditions.

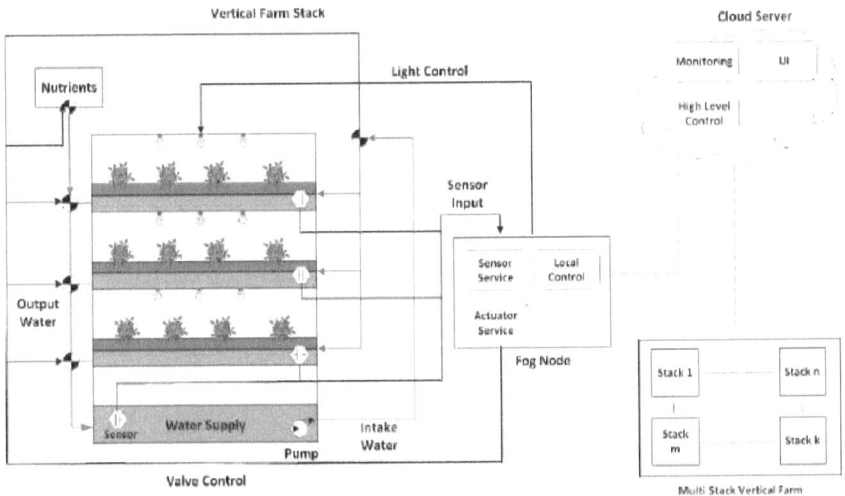

High-level architecture for indoor vertical farming stack. Source: Isakovic et al., 2019

The controlled environment means plants are not affected by seasonal changes or unpredictable weather, ensuring a **steady supply of fresh produce.**

The primary advantage of indoor vertical farming is its **ability to produce crops consistently** throughout the year. This method is particularly **beneficial in regions with extreme climates** or **limited agricultural land**.

Additionally, the use of advanced technologies **allows for precise management** of plant growth, leading to higher yields and better quality produce.

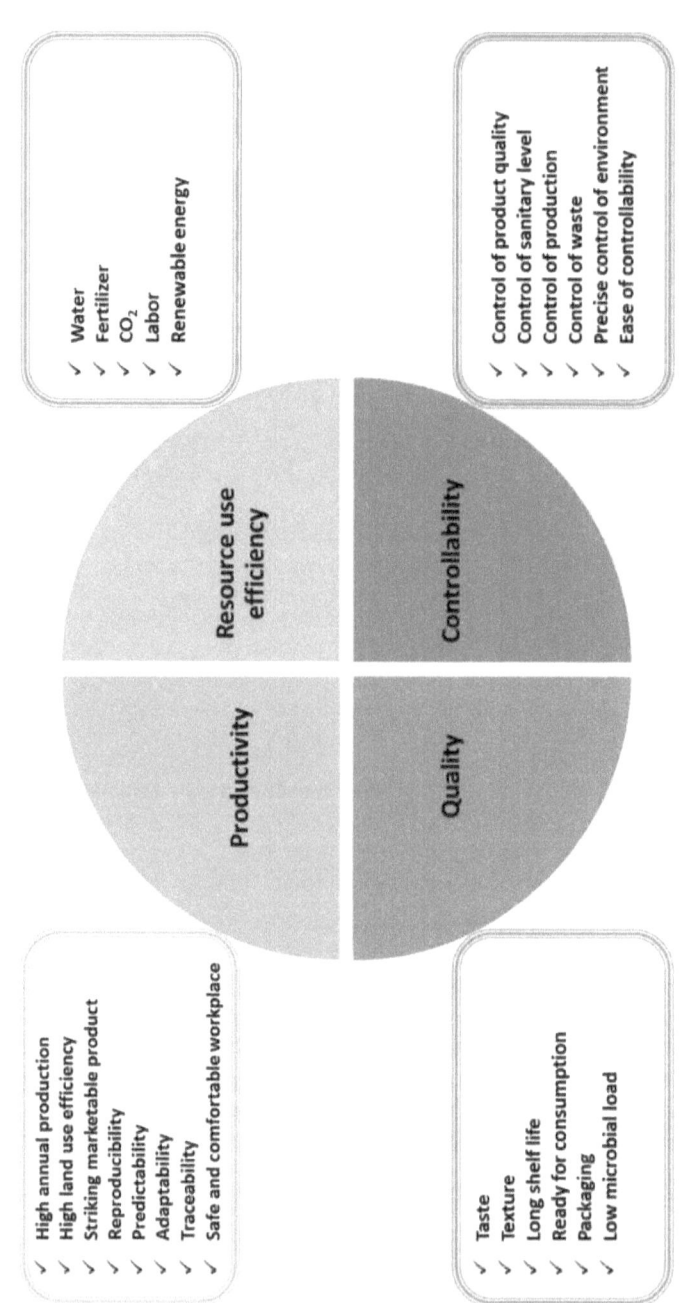

Advantages of production in indoor vertical farming systems. Source: Lastochkina et al., 2022

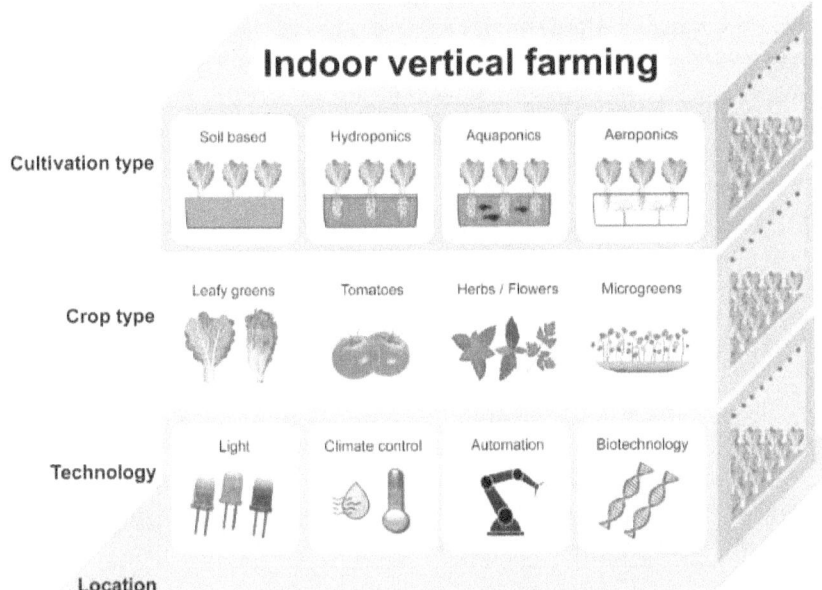

Four main considerations in starting and maintaining an indoor vertical farm. Indoor vertical farming is the practice of producing food in a controlled environment with artificial lighting. This technique aims to maximise crop output in limited spaces independently of weather conditions. The four main considerations in starting and maintaining an indoor vertical farm are cultivation type, crop type, technology and location. Source: Wong *et al.*, 2020

Experimental indoor vertical farming setup. Source: Isakovic *et al.*, 2019

The primary principle of indoor vertical farms is to **utilise internal vertical space efficiently by stacking growing areas**. This is typically achieved through shelving units or towers that extend from floor to ceiling. Each level is equipped with its own lighting, irrigation, and sometimes climate control systems, creating a modular and scalable farming environment.

Most indoor vertical farms utilise **hydroponic or aeroponic systems** for crop cultivation. These soilless growing methods offer precise nutrient delivery and water conservation. Plants are grown in trays, tubes, or other containers that fit within the vertical structure, optimising the use of space.

Many indoor vertical farms **incorporate automation** to enhance efficiency and reduce labour costs. This includes automated systems for seeding, harvesting, and packaging, as well as robots that monitor plant health and move crops through different growth stages. Automation allows for precise control and monitoring, ensuring optimal growing conditions and reducing human error.

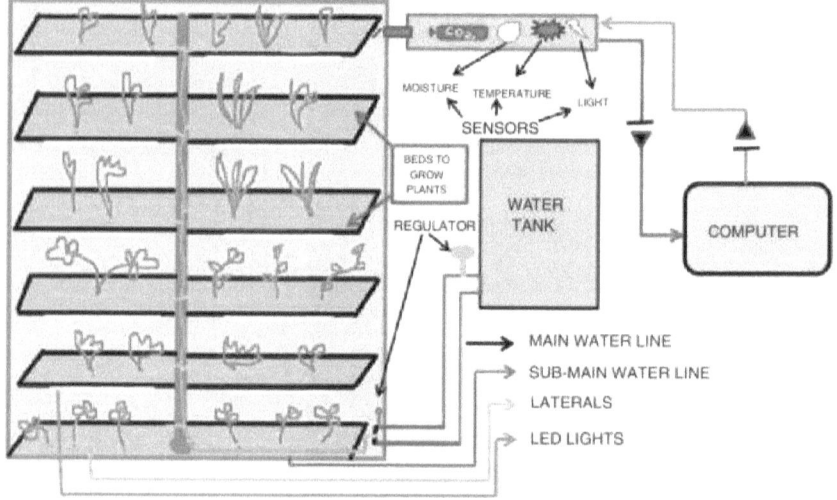

One of the models of automated internal vertical farm design. Source: Mishra et al., 2021

Water usage in indoor vertical farms is typically highly efficient. Water is recirculated through hydroponic or aeroponic systems, minimising waste and reducing the need for freshwater.

Some facilities **incorporate water treatment and filtration systems** to maintain water quality and further reduce waste.

Scientific Report

Improving water use efficiency in vertical farming: Effects of growing systems, far-red radiation and planting density on lettuce cultivation
Country: Italy
Publication Date: 23 May 2023
Main focus: This study investigates methods to improve water use efficiency (WUE) in vertical farming, specifically for lettuce cultivation. The research examines the impact of different growing systems (ebb-and-flow and aeroponics), the inclusion of far-red radiation in lighting, and varying planting densities.

Key findings:
- The aeroponic system demonstrated significantly higher WUE, achieving up to 161.9 g FW/L H_2O, compared to 86.0 g FW/L H_2O for the ebb-and-flow system.
- By recovering water from the HVAC system, water use decreased by 67%, while WUE increased by 206%.
- Impact of Far-Red Radiation: Introducing far-red radiation into the lighting spectrum increased WUE by 46–64% without affecting the dry matter content of the crops.
- Higher planting densities, especially up to 733 plants/m², significantly improved yield and WUE, showing an increase of up to 133%.
- The study underscores the potential for innovative technologies and cultivation techniques, such as optimised lighting and planting densities, to greatly enhance resource efficiency in vertical farming systems.

Reference: Carotti, L., Pistillo, A., Zauli, I., Meneghello, D., Martin, M., Pennisi, G., Gianquinto, G., & Orsini, F. Improving water use efficiency in vertical farming: Effects of growing systems, far-red radiation and planting density on lettuce cultivation. *Agr. Water Management* **2023**, 285, 108365. DOI: 10.1016/j.agwat.2023.108365

Crop selection for indoor vertical farms often focuses on **high-value, fast-growing plants**. Leafy greens, herbs, and certain fruiting crops like tomatoes and strawberries are common choices due to their **short growth cycles and high market demand**. These crops can fully exploit the controlled environment to achieve optimal growth and yield.

Indoor vertical farms rely on preventive measures for **pest and disease management**. The controlled environment and absence of soil reduce the risk of many common agricultural pests and diseases. Some facilities use positive air pressure or air filtration systems to further prevent the entry of pests or pathogens, reducing the need for chemical treatments.

Energy use is a significant factor in indoor vertical farms, particularly due to the artificial lighting and climate control systems that require substantial amounts of electricity.

As a result, **energy efficiency and the incorporation of renewable energy sources** are critical for the economic and environmental sustainability of these operations.

Data collection and analysis play a crucial role in **optimising indoor vertical farms**. Sensors monitor various environmental factors, such as temperature, humidity, light levels, and plant growth metrics.

This data allows farmers to fine-tune growing conditions, predict yields accurately, and continuously improve the efficiency of their operations.

> While indoor vertical farms offer numerous advantages, they also face challenges. **High initial setup costs, ongoing energy expenses,** and **limitations on the types of crops** that can be grown profitably are common issues.

One example of indoor vertical farms could be **shipping container farms** (also known as Containerized Vertical Farming). They represent an innovative approach to vertical farming, utilising **repurposed shipping containers** to create controlled environments for growing plants. These compact, modular systems blend traditional agriculture with modern technology to maximise productivity and efficiency within a limited space.

> Shipping container farms can **range in size and complexity**, typically involving the conversion of standard 20-foot or 40-foot shipping containers into self-contained growing environments. These farms are versatile and can be established in urban areas, on school grounds, at restaurants, or in rural settings seeking innovative farming solutions and educational opportunities.

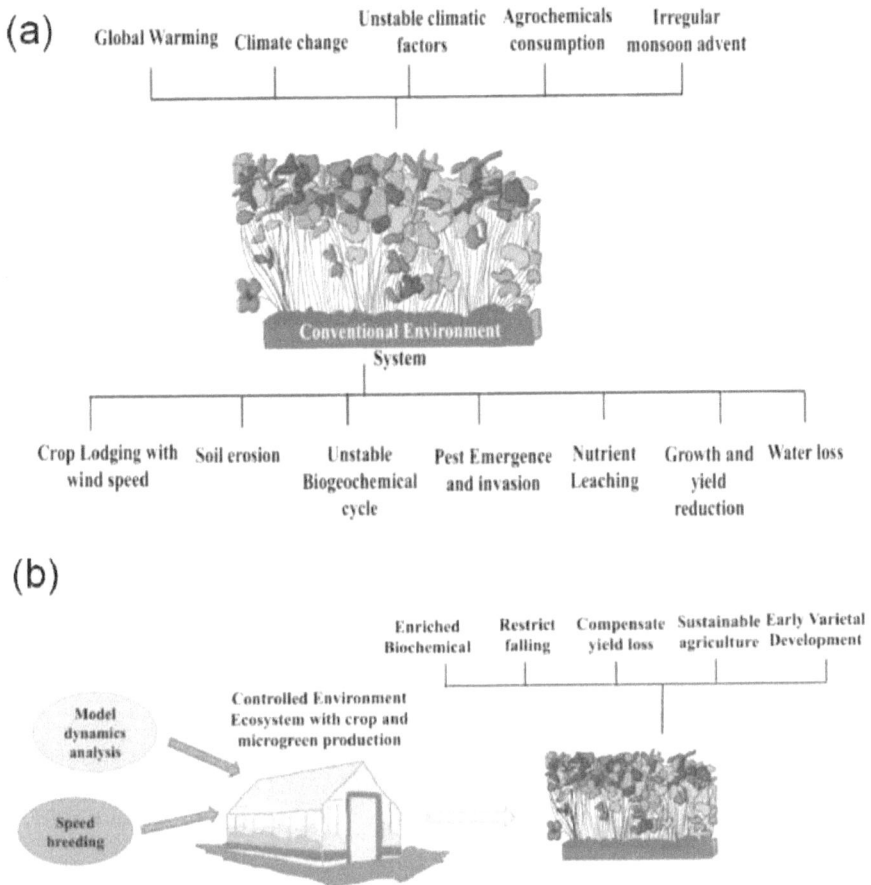

Stability of the ecosystem in conventional and controlled environment agriculture systems.
(a) Conventional Environment Ecosystem. (b) Controlled Environment Ecosystem.
Source: Sharma et al., 2024

The basic structure of a shipping container farm includes several important elements:
1. **insulated walls** to maintain consistent temperatures,
2. **climate control systems** to regulate the internal environment,
3. **LED lighting** to provide the necessary light spectrum for plant growth, and
4. **hydroponic or aeroponic growing systems** for nutrient delivery.

More advanced setups may also feature **solar panels** for energy efficiency, **water recycling systems** to conserve resources, and **sensors** to monitor and adjust growing conditions in real time.

The **choice of plants** for shipping container farms depends on several factors, including the size of the container, desired yield, and specific goals of the farm. **Leafy greens** such as lettuce, kale, spinach, and arugula are common choices due to their rapid growth and high nutritional value. **Herbs** like basil, mint, cilantro, and parsley are also popular, given their high market value and relatively short growth cycles.

At the same time, studies confirmed that **lettuce in shipping containers could not be viable** without efficiency improvements. Romaine lettuce and basil grown in freight containers **couldn't compete** in Europe, but better space and plant density could cut basil costs from €19/kg to €10/kg [Bafort *et al.*, 2022]

Additionally, **microgreens**, which are young vegetable greens harvested shortly after sprouting, thrive in these environments and are highly nutritious.

> Shipping container farms offer several advantages that make them an attractive option for urban agriculture. One of the primary benefits is the ability to **create highly controlled growing environments**. By repurposing shipping containers, these **farms can be established almost anywhere**, making use of underutilised spaces in both urban and rural areas. The **modularity of shipping containers** also allows for scalability; additional containers can be added to expand production as needed.

Another significant advantage is the **potential for year-round production**. The controlled environment within a shipping container farm allows for continuous cultivation, regardless of external weather conditions. This means that fresh produce can be available throughout the year, enhancing food security and reducing seasonal dependencies.

Water efficiency is another key benefit. Shipping container farms typically use hydroponic or aeroponic systems, which drastically reduce water usage compared to traditional soil-based farming methods. These systems recirculate water, minimising waste and conserving this precious resource.

Locating farms closer to urban centres can also **reduce transportation costs** and **emissions**. Fresh produce grown in shipping container farms can reach consumers more quickly, ensuring better quality and reducing the environmental impact of long-distance transportation.

From a social perspective, shipping container farms offer valuable **educational opportunities** and serve as tools for **community engagement**. Schools and community centres can use these farms to teach students and residents about sustainable agriculture practices, fostering a deeper understanding of food systems and environmental stewardship. Additionally, partnerships with local restaurants or markets can strengthen the local economy by providing fresh, locally grown produce.

> Despite their numerous advantages, shipping container farms face several challenges. **One of the main limitations is space**. The compact size of shipping containers restricts the types and quantities of crops that can be grown, potentially limiting the diversity and scale of production.

Another challenge is the **initial investment** required to convert shipping containers into functional farms. The cost of equipment and technology needed for climate control, lighting, and automated systems can be substantial, posing a barrier to entry for some potential farmers.

Maintaining optimal growing conditions within the confined space of a shipping container can also be technically demanding and energy-intensive. Effective climate control is crucial to prevent issues such as overheating or inadequate humidity, which can affect plant health and yield.

As urban populations continue to grow and the demand for sustainable food production increases, shipping container farms are likely to become **more prevalent in cities and towns**. Advances in technology, such as improved climate control systems and more efficient energy use, will enhance the viability and scalability of these farms, making them an even more attractive option for urban agriculture.

<u>Outdoor vertical gardens</u> take advantage of natural elements by utilising external building surfaces such as walls, facades, and rooftops.

This approach relies **primarily on natural sunlight** and often **incorporates rainwater harvesting**. While outdoor systems are more susceptible to seasonal changes and weather variations, they generally have **lower energy costs** and provide additional benefits to urban environments, such as improved building insulation and mitigation of the urban heat island effect.

Moreover, the presence of plants on building exteriors can **improve insulation**, **reduce energy costs** for heating and cooling, and help **combat the urban heat island effect**, where cities become significantly warmer than surrounding rural areas.

Outdoor vertical farming is a good way to **integrate greenery into urban settings**, enhancing the aesthetic appeal and environmental quality of cities. By using available surfaces, such as the sides of buildings or rooftops, this method maximises space efficiency without requiring additional land.

Scientific Report

Study on Outdoor Urban Farming Planters Through Daylight Simulations: A Full-Scale Experiment in Singapore
Country: Singapore
Publication Date: June 2023
Main focus: The study investigates the feasibility and effectiveness of using outdoor urban farming planters in Singapore, focusing on optimising Photosynthetic Photon Flux Density (PPFD) and Daily Light Integral (DLI) for growing Asian greens.

Key findings: The study found that the highest PPFD recorded was 1540 µMol/m²/s at 1300 hrs, and the highest DLI reached 30 mol/m² per day, indicating sufficient light conditions for optimal crop growth. The research suggests that integrating such planters can help reduce reliance on imported food, supporting Singapore's "30 by 30" vision of producing 30% of its nutritional needs locally by 2030.

Reference: Soh, C. B., Haridarshan, R., Chien, S.-C., An, H., Saha, A., & Teoh, M. T. Study on Outdoor Urban Farming Planters Through Daylight Simulations: A Full-Scale Experiment in Singapore. *14th Asia Lighting Conference Proceedings* **2023**. SingaporeTech

---

Outdoor vertical farming can include **green walls** and **rooftop gardens**.

Green walls or living walls are outdoor vertical farming systems that are exposed to natural outdoor conditions. These systems utilise exterior building walls, freestanding structures, or specially designed frameworks to grow plants vertically in an outdoor setting.

Plant selection for green walls depends on local climate, sun exposure, and the specific goals of the installation. Plant selection for green walls involves choosing species that are hardy, low-maintenance, and suited to vertical growth, such as ferns, ivy, and succulents. It is possible to grow certain crops on green walls, such as herbs, leafy greens.

Maintenance of green walls can be more challenging than indoor systems due to exposure to weather, pests, and diseases and the need for more frequent watering and pruning.

> Outdoor vertical gardens often serve **multiple purposes beyond food production**. They can help reduce the urban heat island effect by absorbing sunlight and providing natural cooling. They also contribute to air purification and can increase biodiversity in urban areas by providing habitats for insects and birds.

Rooftop gardens are a form of outdoor vertical farming settings that utilises the often-unused space on top of buildings to grow plants. Rooftop gardens can **vary greatly in size and complexity,** ranging from simple container gardens to extensive green roof systems.

The basic structure of a rooftop garden typically includes several layers. A **waterproof membrane** protects the building's roof, followed by a **root barrier, drainage layer, growing medium**, and finally the **plants** themselves. More advanced systems might incorporate irrigation systems, greenhouse structures, or even aquaponic setups.

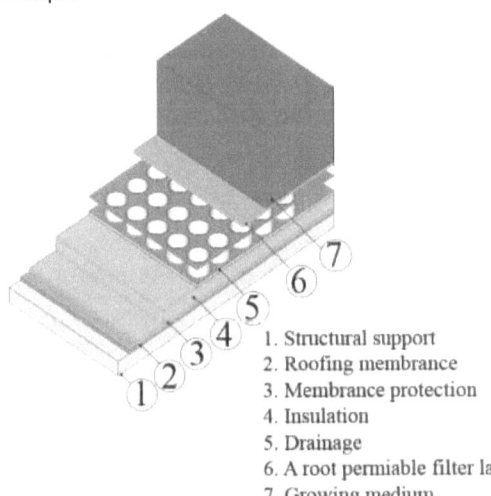

1. Structural support
2. Roofing membrane
3. Membrane protection
4. Insulation
5. Drainage
6. A root permiable filter layer
7. Growing medium

An example of the layers of a green roof. Source: Daneshyar, 2024

Plant selection for rooftop gardens depends on various factors including climate, weight restrictions of the roof, and the garden's purpose. Many rooftop gardens grow a mix of **ornamental** and **edible plants**. **Herbs**, **leafy greens**, and **small fruiting plants** are common choices for food production, while **native plants** or **drought-resistant species** are often used for environmental benefits.

One key advantage of rooftop gardens is their **ability to make use of otherwise unproductive urban space**. They can help increase a city's green cover, contributing to improved air quality and reduced urban heat island effect.

Water management is crucial in rooftop gardens. Many systems incorporate rainwater harvesting and water-efficient irrigation methods. Some advanced designs even use greywater recycling systems from the building below.

Rooftop gardens face unique challenges. **Weight limitations** of the building structure must be carefully considered. **Exposure to wind and intense sunlight** can create harsh growing conditions. Additionally, **access for maintenance** can sometimes be difficult.

Despite these challenges, rooftop gardens offer **numerous benefits**. They can provide fresh, local produce, reduce a building's energy costs through improved insulation, extend the life of the roof by protecting it from UV radiation, and create valuable green spaces in dense urban environments.

> Many rooftop gardens also **serve social functions**. They can be used as community gathering spaces, for educational programs, or as relaxation areas for building occupants. Some restaurants have even begun using rooftop gardens to **supply ultra-fresh ingredients** for their kitchens.

From an environmental perspective, rooftop gardens can play a role in **managing stormwater runoff** in cities, **reducing the burden on municipal drainage systems** during heavy rains.

Scientific Report

Residential Rooftop Urban Agriculture: Architectural Design Recommendations
Country: Cyprus
Publication Date: 25 February 2024

Main focus: This research focuses on proposing design recommendations for architects interested in designing residential buildings capable of rooftop food production.

Key findings:
- Rooftop farming methods can include open-air production, "low-tech" greenhouses, and "high-tech" greenhouses.
- Proper insulation and waterproofing, access to water and electricity, and consideration of load-bearing capacity are essential for successful rooftop farming.
- High-tech greenhouses require 231 times more energy to produce one kilogram of tomatoes compared to open-field farming in Washington State.

Reference: Daneshyar, E. Residential Rooftop Urban Agriculture: Architectural Design Recommendations. *Sustainability* **2024**, *16*, 1881. DOI: 10.3390/su16051881

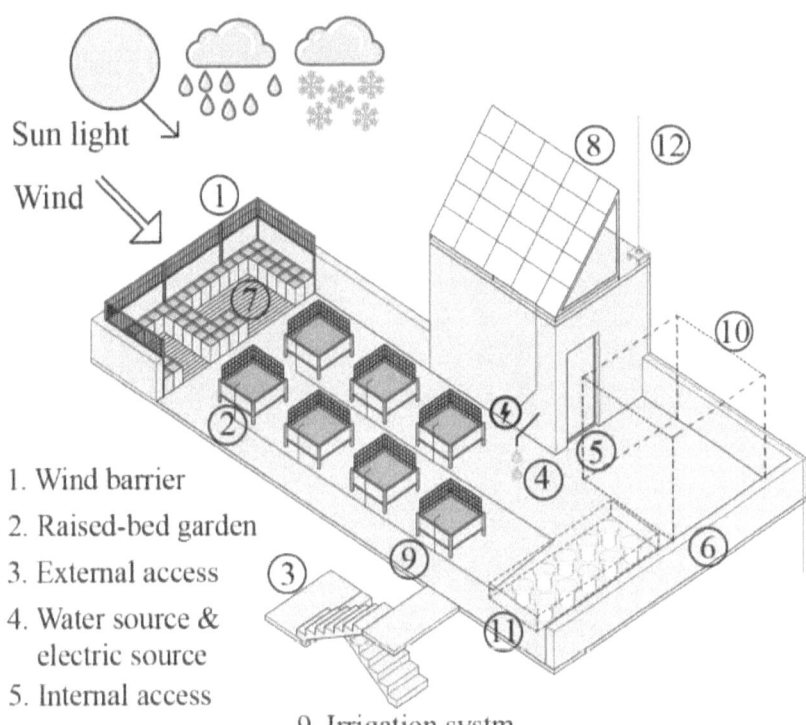

1. Wind barrier
2. Raised-bed garden
3. External access
4. Water source & electric source
5. Internal access
6. Parapet
7. Container garden
8. Solar panel
9. Irrigation systm
10. Reserved area for mechanical system
11. Exhaust vents are grouped together.
12. lightning protection system.

Various factors should be considered before designing a rooftop farm.
Source: Daneshyar, 2024

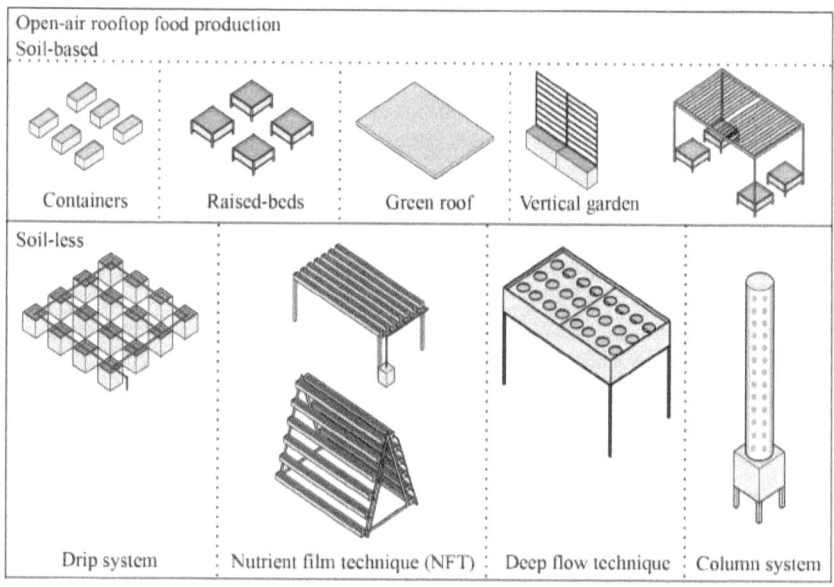

Soil-based and soilless cultivation methods as part of open-air rooftop food production.
Source: Daneshyar, 2024

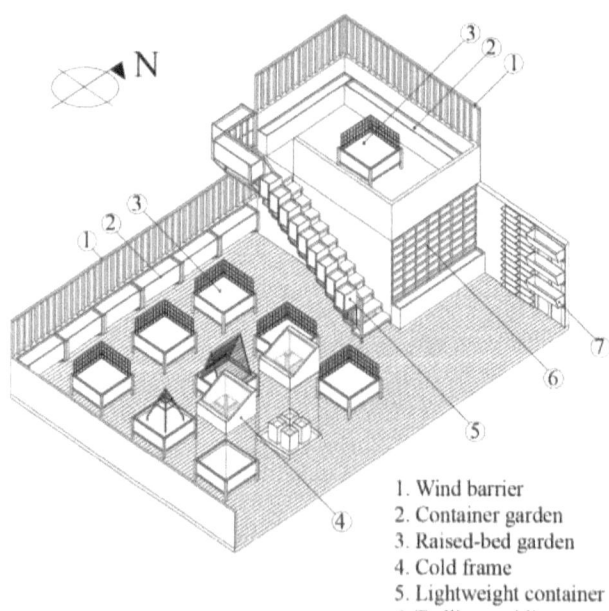

1. Wind barrier
2. Container garden
3. Raised-bed garden
4. Cold frame
5. Lightweight container
6. Trellis providing support
7. Modular wall planters

Design proposal for roof garden. Source: Daneshyar, 2024

Soilless and Substrate-based Growing Methods

The root system could grow **in the air** (aeroponic cultivation), **on the liquid nutrient solution** (hydroponic cultivation), and **on a solid substrate** with added nutrient solution (substrate cultivation) [Birlanga et al., 2022]

Growing medium used (A) and type of farming methods (B) on vertical farming.
*Unspecified mix; include peat moss, perlite, sand, vermiculite, coir, pumice stone, and pine bark. Source: Nájera et al., 2023

In the world of vertical farming, **substrate-based systems** have emerged as a popular and efficient way of cultivation. This approach combines the **space-saving benefits** of vertical farming with the **stability and familiarity** of traditional growing mediums

Substrate-based vertical farming is a method where plants are grown in **vertical stacks** or towers, using a **solid growing medium**. This method provides support for the plant roots and helps deliver nutrients and water efficiently.

Both **organic** and **inorganic** substrates can be used in vertical farming [Birlanga et al., 2022]. Organic substrates include natural materials like peat and by-products of agriculture such as coconut fibre, cereal straw, and wood shavings. Inorganic substrates include naturally porous materials like sand and volcanic gravel, as well as industrially processed materials like rock wool, fibreglass, perlite, and vermiculite.

The choice of substrate can significantly impact plant growth and system efficiency. Common substrates include:

1. **Coconut coir:** Made from coconut husks, this eco-friendly option retains water well and provides good aeration. Good for moisture-loving crops like tomatoes and leafy greens.
2. **Rockwool:** A mineral fibre that offers good water retention and root support. Good for consistent hydration needs of fruit-bearing plants like peppers.
3. **Perlite:** A lightweight, volcanic glass that improves drainage and aeration. Suitable for root vegetables and succulents due to its good drainage.
4. **Vermiculite:** A mineral that expands when heated, providing good water retention and nutrient holding capacity. Good for seed starting and leafy greens with high moisture requirements.
5. **Expanded clay pellets:** These provide good drainage and can be reused multiple times. Preferable for orchids and hydroponic systems.

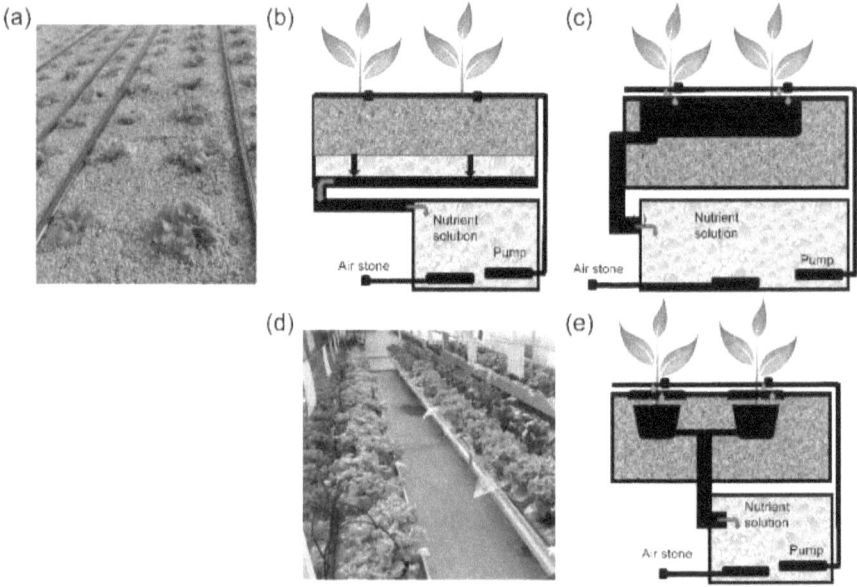

Substrate cultivation systems. (a) Lettuce plants growing in sand as an inert substrate, (b) a scheme of plants growing directly onto substrate, (c) a scheme of plants growing in containers, (d) lettuce plants growing in individual containers, and (e) a scheme of plants growing in individual containers. Source: Birlanga *et al.*, 2022

Substrate cultivation systems provide **better aeration** than water cultivation systems, but they require a continuous **flow of water** to achieve maximum production.

At the same time, vertical farming relies heavily on **soilless growing techniques**. The three main methods used are **hydroponics, aeroponics**, and **aquaponics**. Each of these systems has unique characteristics and advantages, making them suitable for different types of crops and growing conditions.

Hydroponics is the most common method in vertical farming, and has a rich history dating back to the 17th century. Belgian scientist **Jan Baptista van Helmont** (1580 – 1644) partially discovered the process of photosynthesis. He grew a willow tree in a weighted amount of soil. After five years, he discovered that the willow tree weighed about 74 kg more than it did at the start, concluding that the tree grew primarily by absorbing water.

Later hydroponics origins can be traced to the publication of *Sylva Sylvarum* (1627), also known as "A Natural History", by the English scientist **Francis Bacon** (1561 – 1626), which explained the "water culture" and growing terrestrial plants without soil.

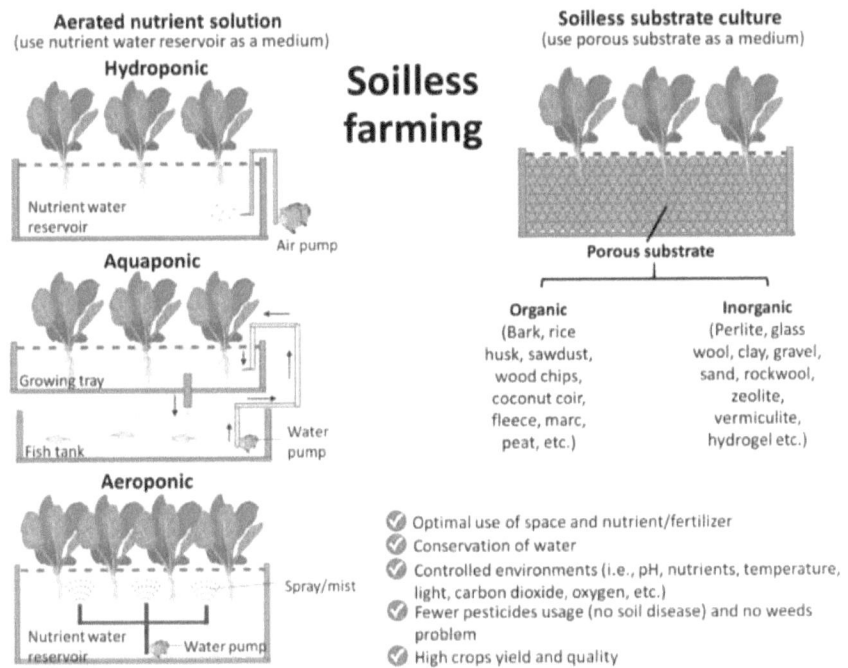

Types of soilless farming and its advantages. Source: Maluin et al., 2021

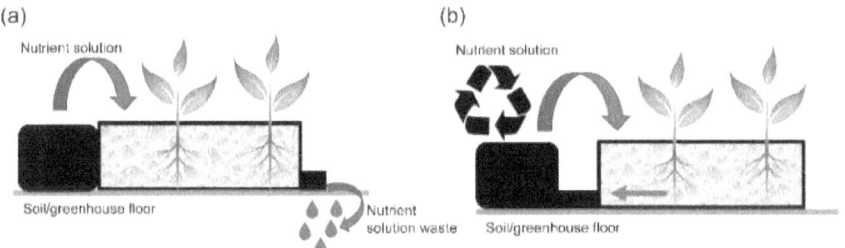

Soilless cropping systems as regards to nutrient solution uses. (**a**) A scheme of an open-loop system in which nutrient solution residues are not recycled, and (**b**) a scheme of a closed-loop system in which nutrient solution residues are reintroduced into the system. Source: Birlanga et al., 2022

The key characteristic of any hydroponic system is that **plants grow with their roots in nutrient-rich water, so they don't need soil**.
The water is **carefully balanced with essential minerals and elements** that plants need to thrive. Several types of hydroponic systems are widely used.

Nutrient Film Technique (NFT): A thin film of nutrient solution flows over the roots, providing a constant supply of water, oxygen, and nutrients.

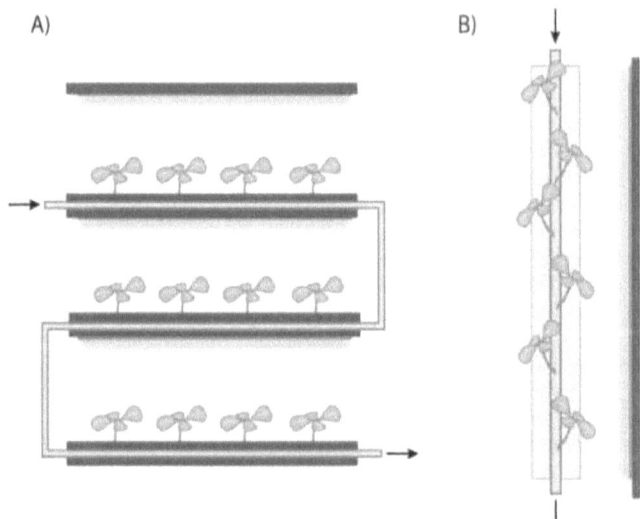

Structure of the hydroponics vertical farm (**A**) Nutrient film technique. Each tray has LED lighting above it. (**B**) Drip irrigation system. The column is dripped with nutrient water. This device can either allow for more natural daylight or be illuminated from the side by LEDs.
Source: Gentry et al., 2019

Deep Water Culture (DWC): Plants are suspended in a solution with their roots submerged in the nutrient-rich water.

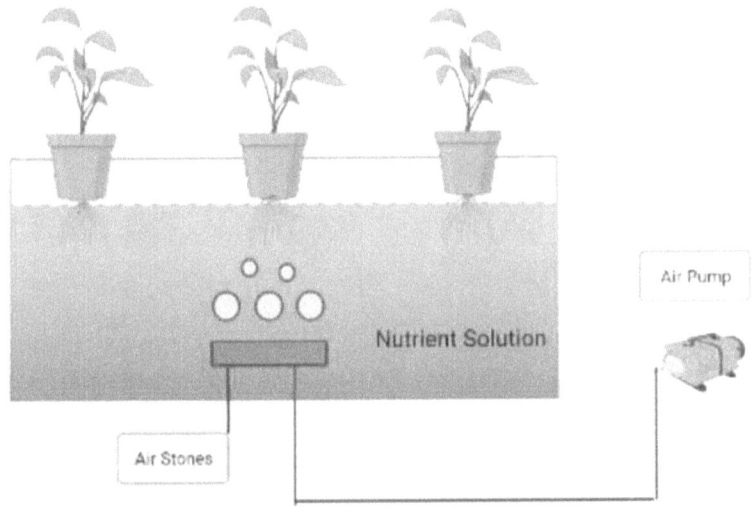

DWC hydroponics system. Source: Rathor *et al.*, 2024

Ebb and Flow Systems: Also known as flood and drain systems, these periodically flood the plant roots with nutrient solution and then drain it away, allowing roots to receive both nutrients and oxygen.

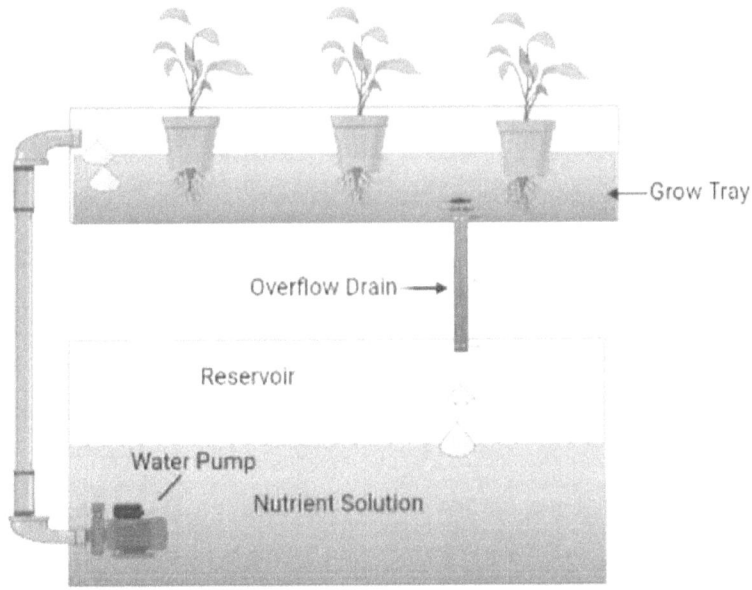

Ebb and flow hydroponics system. Source: Rathor *et al.*, 2024

Wick hydroponics system is a simple, passive method, where nutrient-rich water is drawn up to the plant roots through a wick material, providing necessary nutrients and moisture.

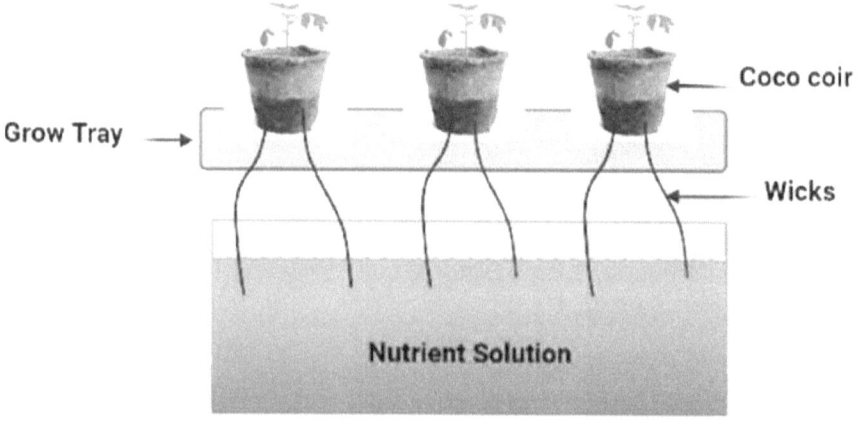

Wick hydroponics system. Source: Rathor *et al.*, 2024

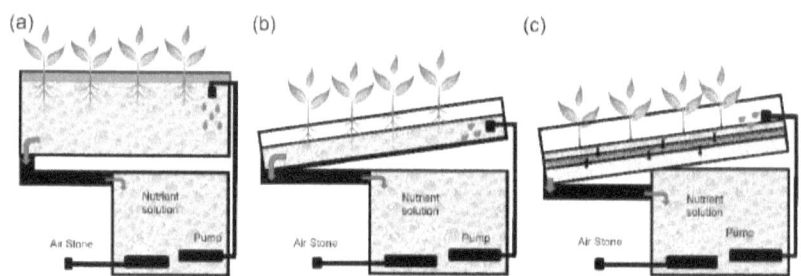

Hydroponic cultivation systems. (a) Deep floating technique (DFT) system, (b) nutrient film technique (NFT) system, and (c) new growing system technique (NGST). Source: Birlanga *et al.*, 2022

Hydroponics allows for **precise control of nutrient delivery**, which can result in **higher yields** and **faster growth**.

It also significantly **reduces water usage** compared to traditional soil-based farming, making it an efficient and sustainable option for growing a wide variety of crops.

When **compared to results** calculated for conventional field production, the hydroponic glasshouse had a 10 times greater yield and 10 times smaller water requirement compared to conventional production. However, the energy demands of the hydroponic glasshouse were around 80 times higher [Beacham et al., 2019]

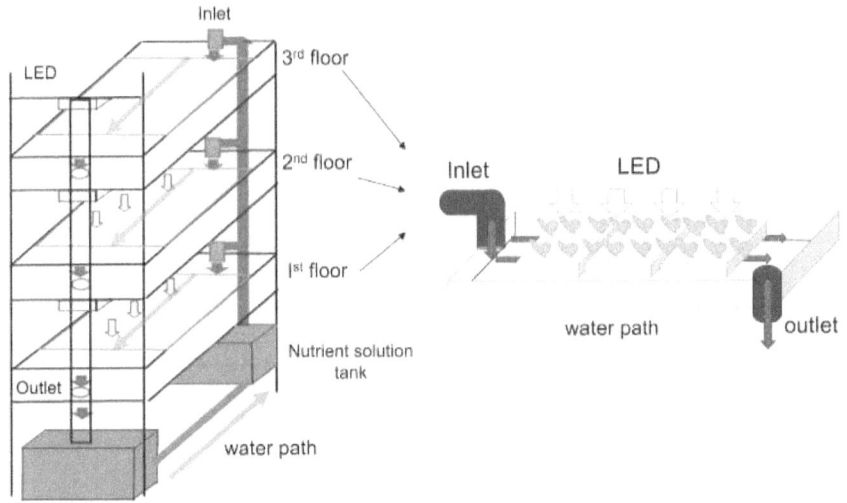

Schematic diagram of the hydroponic indoor farming system for tomatoes.
Source: Lee et al., 2017

Parameters	Average	Min	Max
Atmospheric temperature (°C)	21.7	17.1	24.8
Humidity (%)	61	47	76
Water temperature (°C)	23.2	18.2	27.3
Dissolved oxygen (mg L^{-1})	8.3	7.6	8.8
Light intensity (μmol m^{-2} s^{-1})	198	101	337
pH	7.6	5.8	8.3
EC (mS cm^{-1})	1.07	0.35	1.49
Sampling point (measured time) (n = 45)	Stage I: Day 0, 1, 2, 4, 6, 8, 10, 12 Stage II: Day 14, 16, 18, 20, 22, 24, 27 Stage III: Day 33, 36, 39, 42, 45, 49 Stage IV: Day 52, 56, 59, 63, 66, 69, 72, 76, 79, 83, 86, 90, 93, 97, 100, 104, 107, 111		
Nutrient solution injection time (n = 12)	Stage I: Day 0, 12 Stage II: not supplied Stage III: Day 34, 45 Stage IV: Day 54, 63, 70, 79, 84, 91, 97, 104		

Summary of possible cultivation condition, sampling time, and nutrient solution injection time for tomatoes. Source: Lee et. al., 2017

Advantages and challenges of vertical farming using closed-loop hydroponics. Source: Chiaranunt & White, 2023

Issues	Advantages	Challenges
Water Use	No soil runoff in closed hydroponic systems. Improved water use efficiency.	Production can be constrained by freshwater resources.
Nutrition and Fertilization	Fewer nutrients wasted to runoff. Fine control of nutrient concentrations.	Closed loop systems can increase the risk of nutrient toxicity, if mismanaged.
Disease and pests	Exclusion of pests, pathogens from closed environments. Sanitation of tools, equipment, growing area.	High humidity and temperature may be suitable for pathogens. Rapid spread if pathogen is not excluded.
Crop productivity	Consistent, high yields, depending on the crop.	Major staple crops (rice, wheat, corn) are not feasible to grow in a vertical farm.
Costs	Produce transportation savings and minimization of spoilage. Reduced pesticide requirements.	High setup and operational costs.
Environmental impact	Minimization of fertilizer runoff and downstream eutrophication. Reduced use of synthetic fertilizers and pesticides.	Wastewater accumulation can be high in salts and organic matter. Intensive energy use from LEDs.

In 1997, a NASA-based consortium developed **aeroponics**, a soil-less plant-growth method used in microgravity aboard the Mir space station. Aeroponics takes the concept of **soilless growing** a step further.

In aeroponic systems, **plant roots hang in the air and are sprayed with a nutrient solution.** This method provides roots with **maximum access to oxygen**, which can lead to faster growth rates compared to other methods.

Aeroponics uses even **less water** than hydroponics and allows for easy **inspection of root health**.

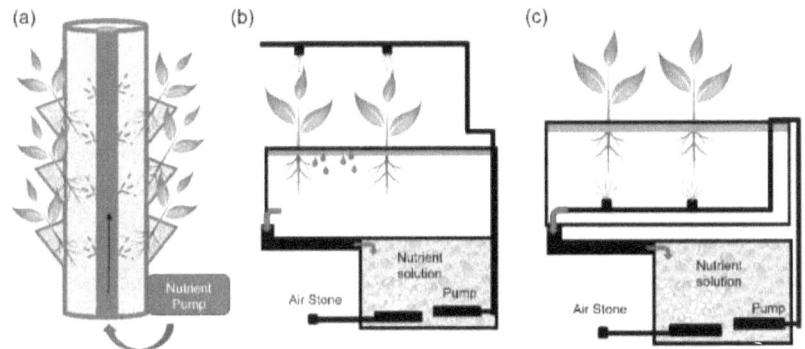

Aeroponic cultivation systems. (a) Spray column system, (b) Schwalbach system, and (c) Aero-Gro system. Source: Birlanga et al., 2022

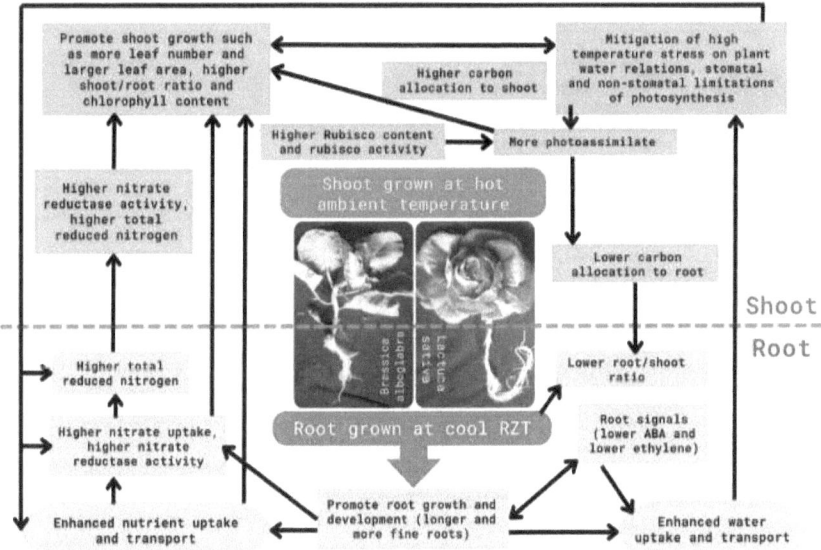

Changes of various physiological parameters of subtropical (*Brassica alboglabra* L.) and temperate vegetable crops (*Lactuca sativa* L.) grown with aeroponic systems by exposing their roots to cooling temperatures (15–25 °C) while their aerial parts were maintained at fluctuating hot ambient temperatures (25–40 °C) in the tropical greenhouse. The changes of different physiological parameters are in comparison to those whole plants grown under hot fluctuating ambient temperatures. Source: He, 2024

However, aeroponic systems require **more technical expertise** to maintain.

> The misting systems **must run consistently** to ensure the plants receive adequate nutrients and moisture. Any malfunction can quickly stress or damage the plants, making **reliable equipment and careful monitoring** essential.

Aquaponics combines **hydroponics with aquaculture** (fish farming). In this symbiotic system, **fish waste provides nutrients for the plants**, while the plants filter the water for the fish.

Beneficial bacteria convert fish waste into forms of nitrogen that plants can use, creating a balanced ecosystem. This method produces both plant crops and fish protein, making it highly efficient in terms of resource use.

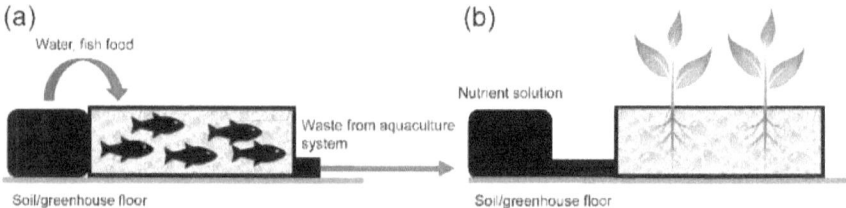

Aquaponic system consists of an (a) aquaculture system for fish production, which is connected to a (b) hydroponic system used for crop production. Source: Birlanga et al., 2022

> Aquaponics requires careful **balancing of fish and plant populations** to maintain system health. It also involves **managing the pH levels and nutrient concentrations** to support both the aquatic life and the plants.

Despite its complexity, aquaponics can be **very sustainable** and offers the added benefit of producing **multiple types of food**.

Each of these methods has its own set of advantages and challenges:

- **Hydroponics** is reliable and scalable, making it well-suited for a wide range of crops. It is the most widely used method in commercial vertical farms due to its versatility and efficiency.
- **Aeroponics** offers rapid plant growth and the least water usage, making it ideal for operations where water conservation is a top priority. It is especially effective for root crops.
- **Aquaponics** provides dual outputs (plants and fish) and promotes sustainability through its closed-loop system. It is appealing for those interested in creating a natural and efficient growing environment.

In all these systems, **maintaining the correct balance of nutrients, pH levels, and oxygen in the water or mist** is crucial. Each method requires careful monitoring and adjustment to ensure optimal plant growth.

The choice between these methods often depends on factors such as the **types of crops being grown**, **available resources**, **technical expertise**, and **specific goals** of the vertical farm.

3. Vertical Farming Layout

This chapter, *Vertical Farming Layout,* comprises three vital components: the crops grown, the environment in which they are cultivated, and the AI-driven systems that manage the interaction between crops and their surroundings.

Crop Selection and Crop Health explores a **diverse range of suitable plants**, including leafy greens, microgreens, herbs, fruits, vegetables, medicinal plants, edible flowers, and cash crops. Understanding the specific requirements of each crop type allows for tailored cultivation strategies that maximise productivity and profitability. Additionally, this section addresses **common pests and diseases**, emphasising the importance of integrated pest management for maintaining crop health and vertical farming system resilience.

Environment explains the critical aspects in creating **ideal growing conditions** within the vertical farm. This encompasses various factors such as lighting, temperature, humidity, air circulation, and carbon dioxide levels.

Alongside these environmental factors, *AI and Internet of Things (IoT)* plays a fundamental role in **monitoring conditions, optimising resource allocation, and streamlining operations** from planting to harvest. By integrating these technologies, vertical farms can achieve higher productivity, consistency, and scalability, enhancing their competitiveness and long-term sustainability.

Crop Selection and Crop Health

Crop selection is crucial in vertical farming, as **not all plants are equally suited** to these systems.

The better adapted plants for vertical farming typically share certain characteristics:

- **Fast growth cycle**: Plants that mature quickly allow for more harvests per year, maximising productivity.
- **Compact size**: Smaller plants use vertical space efficiently.
- **High market value**: To offset the higher production costs of vertical farming, crops should have good profit margins.
- **Low height:** Plants that don't grow too tall are easier to manage in stacked systems.
- **Tolerance to artificial lighting:** Since many vertical farms rely on artificial light, plants should respond well to LED or other artificial light sources.

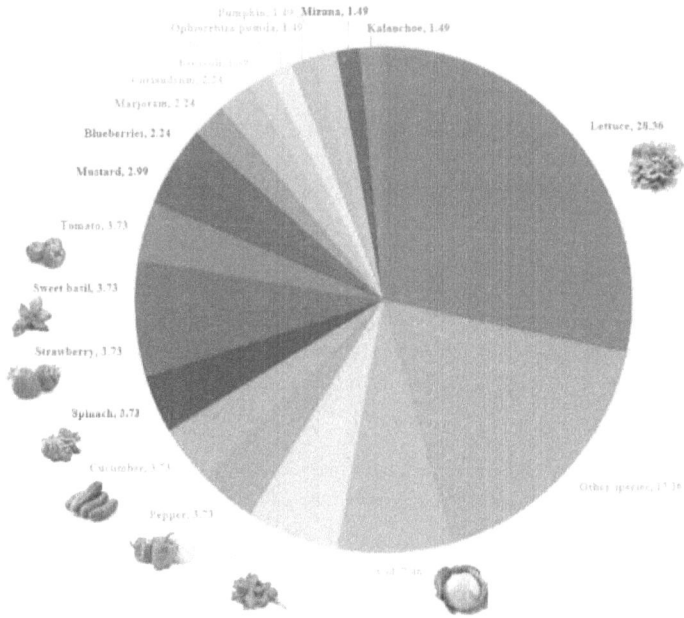

The pie chart shows the relative percentage of published research reported in the literature and used in this review for the different types of vegetables cropped on vertical farming.
Source: Najaro et al., 2023

Based on suitability criteria, the most suitable plants for vertical farming fall into several groups: **leafy greens, microgreens, herbs, fruits, medicinal plants, edible flowers,** and **other cash crops**.

Leafy Greens

1. Lettuce (*Lactuca sativa*)

In vertical farming, lettuce is a popular crop due to its quick growth cycle and variety. Types like romaine, butterhead, and iceberg are cultivated for their crisp texture and mild flavour. Lettuce thrives in controlled environments with precise water and nutrient management, making it ideal for vertical farm setups.

Scientific Report

Utilizing the Power of Plant Growth Promoting Rhizobacteria on Reducing Mineral Fertilizer, Improved Yield, and Nutritional Quality of Batavia Lettuce in a Floating Culture

Country: Turkey
Publication Date: 18 January 2024
Main focus: The study examines the effectiveness of Plant Growth Promoting Rhizobacteria in reducing the use of mineral fertilizers while enhancing the yield and nutritional quality of Batavia lettuce grown in hydroponic floating culture systems.

Key findings: The study found that using PGPR combined with an 80% reduction in mineral fertiliser maintained lettuce yield at 12.3 kg/m², only 5% lower than the 100% mineral fertiliser control. Additionally, the PGPR treatment increased the total phenolic content, flavonoids, and vitamin C levels in lettuce, improving its nutritional quality.

Reference: Ikiz, B., Dasgan, H. Y., & Gruda, N. S. Utilizing the power of plant growth promoting rhizobacteria on reducing mineral fertilizer, improved yield, and nutritional quality of Batavia lettuce in a floating culture. *Scientific Reports* **2024**, 14, 1616. DOI: 10.1038/s41598-024-51818-w

2. Spinach (*Spinacia oleracea*)

Spinach is another leafy green well-suited for vertical farming, known for its nutrient density and tender leaves. It grows rapidly and requires careful control of light and temperature to prevent bolting (premature flowering). Spinach can be grown year-round in vertical farms, providing a consistent supply for salads and cooking.

3. Kale (*Brassica oleracea var. sabellica*)

Kale's hardiness makes it a good candidate for vertical farming. It can withstand a range of temperatures and has a relatively long harvest period. Vertical farms can efficiently produce various kale varieties, including curly and Tuscan kale, which are rich in vitamins and suitable for fresh consumption or cooking.

4. Arugula (*Eruca vesicaria*)

Arugula, known for its peppery flavour, is ideal for vertical farming due to its quick growth and compact size. It requires a carefully controlled environment to maintain its distinctive taste and tender texture. Vertical farms can optimise growing conditions, such as light intensity and nutrient delivery, to enhance the quality of arugula.

5. Swiss Chard (*Beta vulgaris subsp. vulgaris*)

Swiss chard, with its vibrant stems and large leaves, is well-suited to vertical farming systems. It can be grown in high-density configurations, making efficient use of space. Swiss chard requires balanced nutrient solutions and consistent light exposure to develop its characteristic colours and flavours, making it a visually appealing and nutritious crop for vertical farming.

Vertical hydroponic racks with artificial lighting used to grow A) arugula and B) kale under a planting density of 12 individuals per tier over a 28-day growth cycle.
Source: Song *et al.*, 2024

Scientific Report

An IoT Based Drip Irrigation System for Vertical Farming in Rainshelter
Country: India 🇮🇳
Publication Date: 25 May 2023
Main focus: The project focuses on developing an IoT-based automated drip irrigation system for vertical farming within a rain shelter to optimise water usage and improve crop growth of leafy green vegetable amaranth (*Amaranthus spp.*).

Key findings:
- The IoT-based drip irrigation system improved water efficiency by 30% compared to traditional irrigation methods.
- Soil moisture levels were maintained at optimal levels, reducing water usage by 25%.
- Crops grown using this system showed a 20% increase in growth metrics, including height and biomass, compared to those grown with conventional irrigation.
- The system enabled real-time monitoring and control, reducing labour costs by 15% and operational costs by 10%.

Reference: Sharma, R. K., Chauhan, V. S., Gupta, S. D., Patel, P. R. An IoT Based Drip Irrigation System for Vertical Farming in Rainshelter. *Journal of Agricultural Technology*, **2023**, Issue 3. DOI:10.1016/j.jagtec.2023.05.012

Example of a vertical farming setup for growing amaranth in a rain shelter.
Source: Sharma *et al.*, 2023

Microgreens

Microgreens are edible seedlings harvested at about 7-14 days after germination when they have two fully formed cotyledon leaves.

They include a variety of vegetables, herbs, and flowers, and are often more aromatic, colourful, and tender than their mature counterparts.

Microgreens are considered **superfoods** due to their high nutritional value, often containing significantly higher concentrations of essential vitamins and minerals compared to mature plants.

Image on the next page: prospects of microgreens as budding live functional food.
(A) Growth conditions required for microgreens cultivation which includes a variety of substrates like vermiculite (1), cocopeat (2), perlite (3) and vermicompost (4), light (quality, intensity and duration), temperature, humidity and nutrient solution.
(B) Microgreens of different plant species: Wheat, Pearl millet, Sama, Mustard, Rocket, Raddish, Beet and Chickpea.
(C) Numerous "OMICS' approaches such as Genomics, Transcriptomics, Proteomics, Metabolomics, Epigenomics, along with Transgenics, Gene editing and Sequencing based approaches can be integrated with bioinformatics tools and artificial intelligence
(D) to tag Quantitative Trait Loci (eQTLs: mRNA expression Quantitative Trait Loci; meQTLs: Methylation Quantitative Trait Loci; PQTL: Protein Quantitative Trait Loci; sQTL: splicing Quantitative Trait Loci; mQTLs: Metabolic Quantitative Trait Loci) and candidate genes identification for microgreen related desirable traits
(E) like nutrients, flavour, colour, early germination, yield and shelf life. The data generated from Integrated "OMICS' approaches can be further utilized in molecular breeding to produce nutritionally rich varieties with improved shelf life through Marker-Assisted Breeding and producing biofortified microgreens with targeted micro/macro nutrients (like iron, zinc, magnesium, calcium) enrichments.
(F) Biofortification of target microgreen is also possible by agronomic approaches (incubation with microorganisms like Bacillus, Rhizobium, Azotobacter, Pseudomonas indica), nanotechnology (nano-biofortification) and seed priming. Bioavailability of nutrients and minerals of microgreens can be stabilized through pre and post-harvest management strategy
(G). Improved microgreens with desired nutrients can be either consumed fresh as garnishes in soups, sandwiches, salads or processed to develop value-added products (like noodles, breads, drinks, cookies etc.) to overcome nutrient deficiency.
Source: Gupta *et al.*, 2023

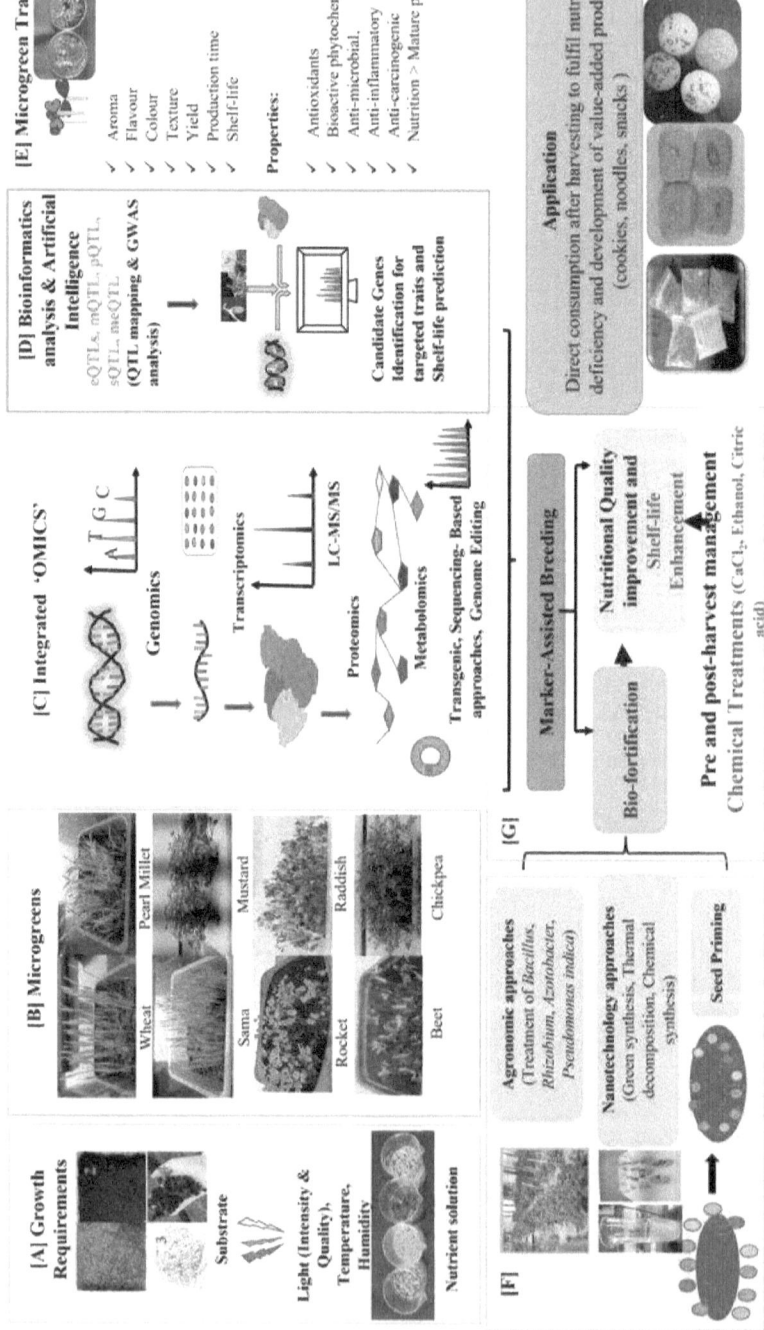

Crop/Plant	Substrate used	Remarks/Findings
Ocimum basilicum L.- Basil, Eruca vesicaria (L.) Cav. subsp. Sativa (Mill.) Thell - Rocket	Hydroponics (Soil-less medium)	High concentrations of some minerals
Eruca sativa Mill. - Rocket, Ocimum basilicum L. - Green Basil, Ocimum basilicum var. Purpurescens - Red basil	Vermiculite, coconut fiber, jute	Substrate significantly regulates nitrate concentration, yield and dry matter percentage
Hairy basil (Ocimum basilicum L.t. var. citratum Back),	Sand, vermicompost, coconut coir dust, sugarcane filter cake, peat	Local organic biomaterials were identified as suitable substitutes to costly peat-based media for cultivating microgreens.
Sweet basil (Ocimum basilicum Linn.),		
Holy basil (Ocimum sanctum Linn.),		
Huanmoo (Dregea volubilis Stapf),		
Sano (Sesbania javanica Mig.),		
Vine spinach (Basella alba Linn.),		
Rat-tailed radish (Raphanus sativus var. caudatus Linn.),		
Leaf mustard (Brassica juncea Czern. & Coss.),		
Kangkong (Ipomoea aquatica Forsk.)		
Krathin (Leucaena leucocephala de Wit),		
Red radish (Raphanussativus) var "Sango"	White sphagnum peat substrate, Coco coir dust	Microgreens grown on these substrates had permissible levels of nitrate content and microbial growth

Different growth media/substrate used in different plants for microgreens cultivation.
Source: Gupta et al., 2023

Scientific Report

Research on the Influence of Some Technological Factors on the Culture of Microgreens in Vertical Farming

Country: Romania
Publication Date: 2023
Main focus: This study investigates the impact of various technological factors, including fertilisation and lighting sources, on the growth and development of microgreens within vertical farming systems.

Key findings:
- The application of biofertilizers significantly enhanced the growth rate and yield of microgreens, with the Bio•Heaven 5 ml/l water treatment resulting in a production increase of 43.91 g per 0.11 m² compared to the unfertilized control.
- Artificial lighting provided a more intensive growth rate than natural light, with coriander microgreens under artificial light showing an increase of 14.02 g per 0.077 m² over natural light conditions.
- Beetroot microgreens grown under artificial light exhibited a growth increase of 9.96 g per 0.077 m² compared to those under natural light.

Reference: Stupariu IM, Balint MV, Poșta Gh. Research on the influence of some technological factors on the culture of microgreens in vertical farming. *Journal of Horticulture, Forestry and Biotechnology*. **2023** ;27(3):135-140. Journal of Horticulture, Forestry and Biotechnology

Preparation of biological material and sowing (original). Source: Stupariu *et al.*, 2023

Aspects of comparative cultures (original). Source: Stupariu *et al.*, 2023

Average values of the morphological characteristics analyzed in radish cv. 'Sangria' in 4 experimental series. Source: Stupariu *et al.*, 2023

Variant	The height of the plant (cm)						Microgreens production (g per 0,11 m²)	
	On the day 4		On the day 6		On the day 7			
	average	Difference	average	Difference	average	Difference	average	Difference
V₁ – Control (unfertilized)	2.86	0.00	5.02	0.00	5.54	0.00	122.62	0.00
V₂ – Bio·Heaven 2 ml / l apă	3.36	+0.50	5.93	+0.91	7.13	+1.59	146.54	+23.92
V₃ – Bio·Heaven 5 ml / l apă	3.51	+0.65	6.51	+1.49	7.57	+2.03	166.53	+43.91

Mean values of the morphological characteristics analyzed in coriander in 4 experimental series. Source: Stupariu *et al.*, 2023

Variant	The height of the plant (cm)						Microgreens production (g per 0,077 m²)	
	On the day 6		On the day 9		On the day 15			
	average	Difference	average	Difference	average	Difference	average	Difference
V₁ – natural light	1.01	0.50	3.51	0.00	6.01	0.00	45.11	0.00
V₂ – artificial light	1.02	+0.01	4.53	+1.02	7.55	+1.54	59.13	+14.02

Average values of the morphological characters analysed in beetroot cv. 'Bull's Blood' in 4 experimental series. Source: Stupariu *et al.*, 2023

Variant	The height of the plant (cm)						Microgreens production (g per 0,077 m²)	
	On the day 6		On the day 9		On the day 15			
	average	Difference	average	Difference	average	Difference	average	Difference
V₁ – natural light	1.03	0.00	4.03	0.00	6.52	0.00	31.15	0.00
V₂ – artificial light	1.05	+0.02	4.52	+0.49	8.21	+1.69	41.11	+9.96

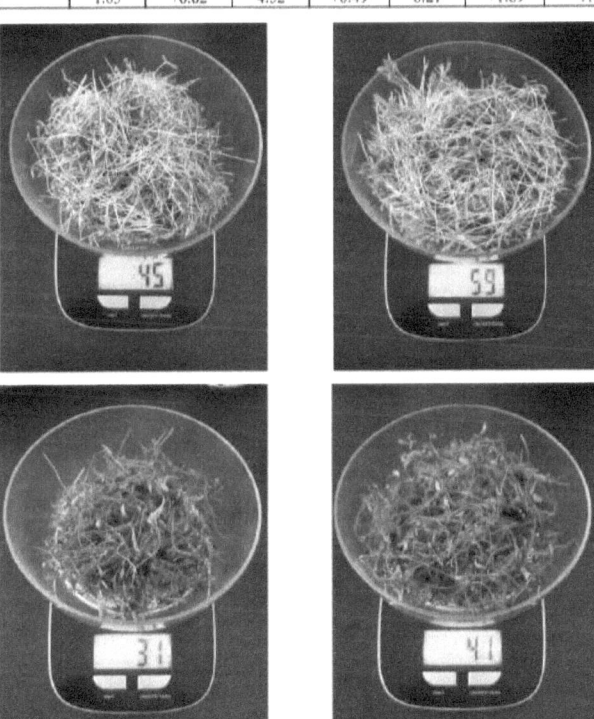

Aspects from the quantitative determinations (original). Source: Stupariu *et al.*, 2023

Both leafy greens and microgreens are ideal because they **grow quickly**, **have a small footprint**, and **command good market prices**, especially when sold as "locally grown" or "pesticide-free"

Herbs

Herbs are good choices due to their high value, compact size, and strong market demand for fresh, locally-grown varieties.

Suitable herbs for hydroponics. Source: Ghazal et al., 2023

Common Name	Latin Name
Tarragon	Artemisia dracunculus L.
Peppermint	Mentha x piperita L.
Green Mint	Mentha L. [such as: Mentha x piperita L. Mentha spicata L. Mentha pulegium L.]
Oregano	Origanum vulgare L.
Basil	Ocimum basilicum L.
Sage	Salvia officinalis L.
Stevia	Stevia rebaudiana (Bertoni) Hemsl.
Lemon Balm	Melissa officinalis L.
Rosemary	Rosmarinus officinalis L.

Scientific Report

Morphological and Physiological Responses in Basil and Brassica Species to Different Proportions of Red, Blue, and Green Wavelengths in Indoor Vertical Farming

Country: USA

Publication Date: 2020

Main focus: The study investigates the effects of different proportions of red, blue, and green wavelengths on the growth and physiological responses of basil and Brassica species in indoor vertical farming systems.

Key findings:
- The study found that total phenolics content in green basil was highest under 76% red and 24% blue light, measuring 24 mg/plant.
- In purple basil, total anthocyanin content was highest under 76% red and 24% blue light at 47 mg/plant.
- Green kale exhibited the highest total flavonoid content under 44% red, 12% blue, and 44% green light, with 52 mg/plant .

Reference: Dou, H., Niu, G., Gu, M., & Masabni, J. Morphological and Physiological Responses in Basil and Brassica Species to Different Proportions of Red, Blue, and Green Wavelengths in Indoor Vertical Farming. *Journal of the American Society for Horticultural Science J. Amer. Soc. Hort. Sci.* **2020**, *145*(4), 267-278. Retrieved Aug 4, 2024, from DOI: 10.21273/JASHS04927-20

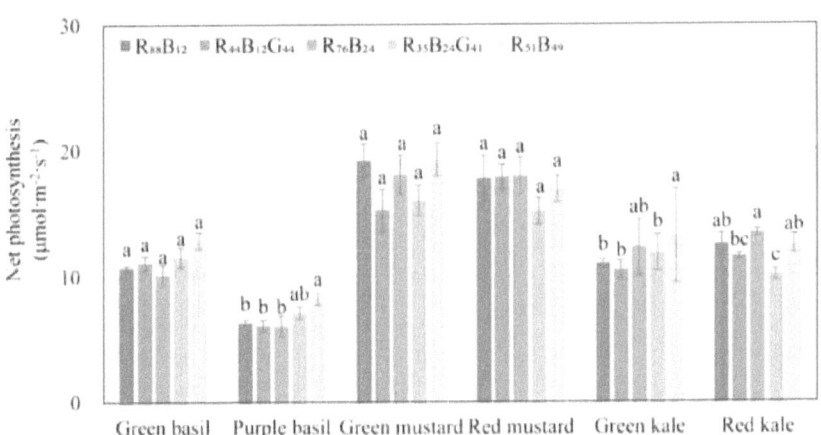

Net photosynthesis of green basil 'Improved Genovese Compact', purple basil 'Red Rubin', green mustard 'Amara', red mustard 'Red Giant', green kale 'Siberian', and red kale 'Scarlet' grown under five light quality treatments: three combined red (R) and blue (B) light combinations, R88B12 (the proportions of R and B wavelengths were 88% and 12%, respectively), R76B24 (the proportions of R and B wavelength were 76% and 24%, respectively), and R51B49 (the proportions of R and B wavelengths were 51% and 49%, respectively) and two white light combinations, R44B12G44 [white light-emitting diode with R, B, and green (G) wavelength proportions of 44%, 12%, and 44%, respectively] and R35B24G41 (white fluorescent light with R, B, and G wavelength proportions of 35%, 24%, and 41%, respectively). Means followed by different lowercase letters indicate a significant difference according to one-way analysis of variance (P # 0.05). Source: *Dou et al., 2020*

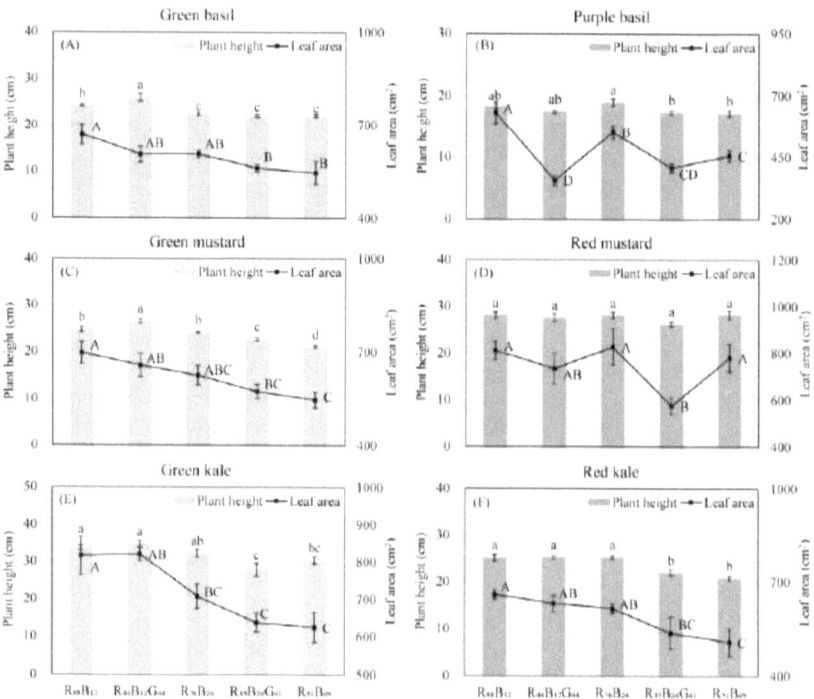

Plant height and leaf area of (A) green basil 'Improved Genovese Compact', (B) purple basil 'Red Rubin', (C) green mustard 'Amara', (D) red mustard 'Red Giant', (E) green kale 'Siberian', and (F) red kale 'Scarlet' grown under three combined red (R) and blue (B) light combinations, R88B12 (the proportions of R and B wavelengths were 88% and 12%, respectively), R76B24 (the proportions of R and B wavelengths were 76% and 24%, respectively), and R51B49 (the proportions of R and B wavelengths were 51% and 49%, respectively) and two white light combinations, R44B12G44 [white light-emitting diode with R, B, and green (G) wavelength proportions of 44%, 12%, and 44%, respectively] and R35B24G41 (white fluorescent light with R, B, and G wavelength proportions of 35%, 24%, and 41%, respectively). Means followed by different lowercase and uppercase letters indicate a significant difference for plant height and leaf area, respectively, according to one-way analysis of variance (P # 0.05). Source: *Dou et al., 2020*

Fruits

1. Tomatoes (*Solanum lycopersicum*)

In vertical farming, tomatoes are a favoured crop due to their high market value and variety. They require support structures and careful environmental control to manage factors like light, humidity, and temperature, which are crucial for fruit set and quality. Vertical farms can grow various types of tomatoes, including cherry, grape, and beefsteak, providing fresh produce year-round.

For example, it has been experimentally proven that tomatoes can grow in many different environments. They do well in temperatures between 10 and 35 degrees Celsius.

The best humidity for their growth, flowering, and fruit development is **between 30% and 90%**. When grown hydroponically, the nutrient solutions for tomatoes should have an electrical conductivity **between 0.8 and 2.5 millisiemens per centimetre** and a pH level **between 5 and 7** [Lee et al., 2017].

Scientific Report

Growth, Fruit Yield, and Bioactive Compounds of Cherry Tomato in Response to Specific White-Based Full-Spectrum Supplemental LED Lighting
Country: South Korea
Publication Date: 9 April 2022
Main focus: This study investigates the effects of specific white-based full-spectrum supplemental LED lighting on the vegetative and reproductive growth, fruit yield, and bioactive compound content in cherry tomatoes grown in a greenhouse setting.

Key findings:
The study found that the specific full-spectrum LED treatments (SFL1 and SFL2) resulted in significant improvements in various growth parameters compared to control and other lighting treatments:
- Shoot fresh weight increased by approximately 50% and 60%, respectively.
- Leaf area increased by 38.2% and 45.2%.
- Total fruit yield increased by 73.1% and 70.7%.
- Energy use efficiency (EUE) was highest in SFL1 and SFL2 treatments, recorded at 6.08 g FW kWh^{-1} and 6.00 g FW kWh^{-1}, respectively.
- Bioactive compounds, such as chlorogenic acid and quercetin, showed significant increases under SFL treatments, with chlorogenic acid content reaching up to 78.93 g plant^{-1} and quercetin up to 1036.47 g plant^{-1}

Reference: Nguyen, T.K.L.; Cho, K.M.; Lee, H.-Y.; Sim, H.-S.; Kim, J.-H.; Son, K.-H. Growth, Fruit Yield, and Bioactive Compounds of Cherry Tomato in Response to Specific White-Based Full-Spectrum Supplemental LED Lighting. *Horticulturae* **2022**, *8*, 319. DOI: 10.3390/horticulturae8040319

Nutrient solution components for cherry tomatoes used in this study. Source: Nguyen et al., 2022

Type	Chemical	Amount (g 1000 L−1)
A	KNO3	20,200
	Ca(NO3)2·4H2O	35,400
	Fe-EDTA	1500
B	KNO3	20,200
	NH4H2PO4	7600
	MgSO4·7H2O	24,600
	H3BO3	114
	MnSO4·4H2O	81
	ZnSO4·7H2O	9
	CuSO4·5H2O	4
	Na2MoO4·2H2O	1

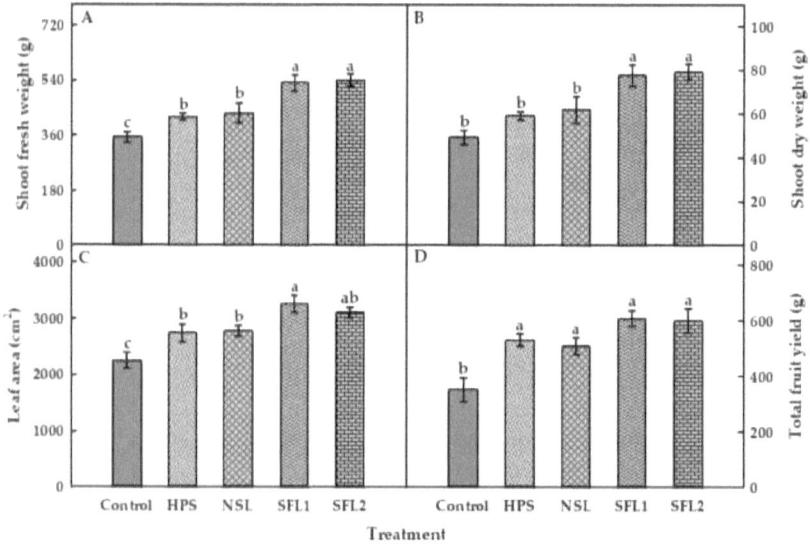

Shoot fresh (**A**) and dry (**B**) weights, leaf area (**C**), and total fruit yield (**D**) of cherry tomato plants grown in different SL sources after 7 weeks of treatment. Control, natural light; HPS, high-pressure sodium lamps; NSL, narrow-spectrum LEDs; SFL, specific full-spectrum or W light sources. Different letters above the bars indicate significant difference at $p < 0.05$ ($n = 6$). Each value in the figure is the mean of 6 replicates (per plant) for each SL source. Total fruit yield was calculated with 5 clusters at each plant. Source: Nguyen et al., 2022

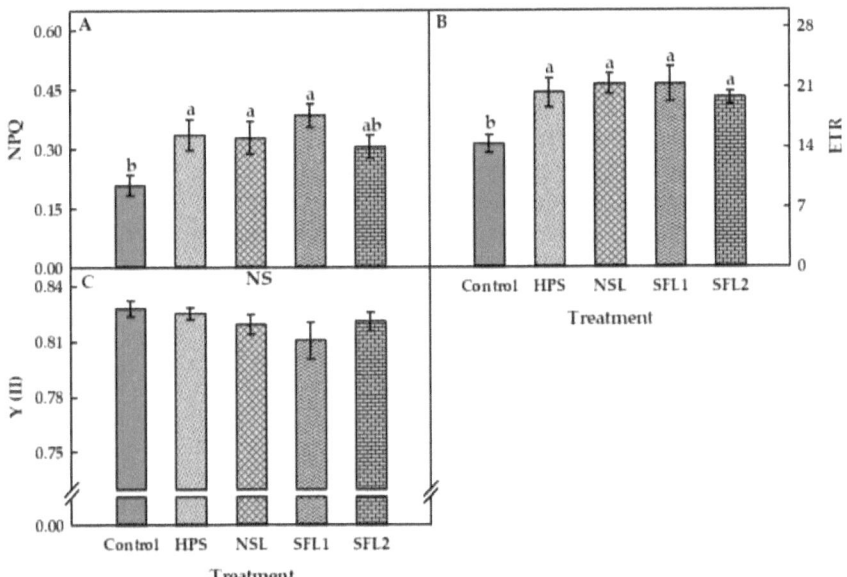

NPQ (**A**), ETR (**B**), and Y (II) (**C**) of cherry tomato plants grown with different SL sources after 7 weeks of treatment. Control, natural light; HPS, high-pressure sodium lamps; NSL, narrow-spectrum LEDs; SFL, specific full-spectrum or W light sources. Different letters above the bars indicate a significant difference at $p < 0.05$ ($n = 5$). NS; not significant. Source: Nguyen et al., 2022

2. Strawberries (*Fragaria × ananassa*)

Strawberries thrive in vertical farming systems, where their need for well-drained soil and controlled temperature can be easily managed. Hydroponic setups are particularly effective for strawberries, allowing for precise nutrient delivery and optimal growing conditions. Vertical farms can produce high-quality, flavorful strawberries with a longer growing season than traditional methods.

Scientific Report

The Added Value of Indoor Products: The Strawberry Case
Country: Netherlands
Publication Date: 2024
Main focus: This study explores the potential benefits of producing strawberries in fully controlled indoor environments, examining the impact of stable versus fluctuating climate conditions on yield and quality.

Key findings:
- Strawberries grown in indoor environments under stable climate conditions produced 1.8 times higher yields compared to greenhouse cultivation.

- The high blue light treatment did not enhance nutritional quality and had a negative impact on production due to compact plant architecture and reduced light interception.
- Stable climate conditions resulted in a 4% higher yield compared to fluctuating conditions, with 7% more Class 1 strawberries.
- Organoleptic qualities such as Brix (sugar content) were higher under stable climate and reference light conditions.

Reference: Carpineti, C., Meinen, E., Vanacore, L., Leman, A., Barbagli, T., Ketel, E., van Hoogdalem, M., & Janse, J. The added value of indoor products: the strawberry case. *Wageningen University & Research* **2022**. DOI: 10.18174/657739

Bottom layer with reference treatment (left) and top layer with high blue treatment (right). Source: Carpineti *et al.*, 2024

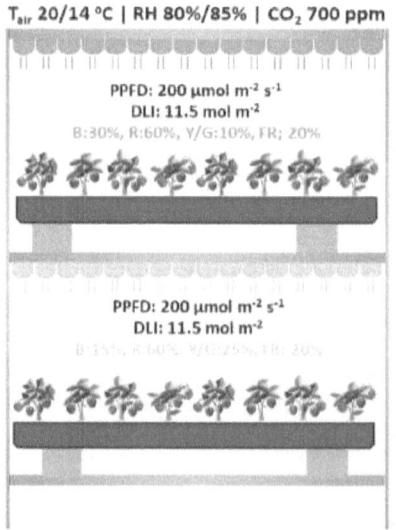 

Schematic overview (left) and picture (right) of the experimental set-up of the Stable treatment. Source: Carpineti *et al.*, 2024

Climate treatment	Light treatment	#leaves	Petiole length (cm)	Calculated leaf area (cm²)	Crop height (cm)	Crop width (cm)	Truss length (cm)
Stable	Reference	20,9±2,1 a	14,4±0,4 a	42,6±1,2 a	25,4±0,3 a	56,6±1,2 a	24,3±0,8 a
Stable	High Blue	18,0±1,3 a	11,7±0,4 b	35,2±1,9 c	20,2±0,6 c	46,9±1,3 c	21,1±1,1 a
Fluctuation	Reference	16,3±1,2 a	12,9±0,4 ab	40,6±1,4 ab	22,8±0,5 b	51,3±1,0 bc	23,5±0,9 a
Fluctuation	High Blue	18,1±1,5 a	12,7±0,3 ab	38,1±1,2 ab	22,1±0,8 bc	52,0±1,2 ab	20,9±0,8 a

Plants morphology results reported as mean values of measurements taken on WOT 5, 7, 9, 11, 13, 15, 19, 20 ± SEM and significance letters with Tukey test (n=6). Source: Carpineti *et al.*, 2024

Climate treatment	Light treatment	Class 1 (Kg/m²)	Class 2 (Kg/m²)	Mildew (Kg/m²)	Total production (Kg/m²)	Not sellable (Kg/m²)
Stable	Reference	7,7	1,9	2,0	11,6	0,7
Stable	High Blue	7,3	1,4	1,8	10,5	0,5
Fluctuation	Reference	7,1	1,6	2,4	11,1	0,7
Fluctuation	High Blue	6,9	1,6	1,8	10,3	0,6

Total fresh yield per class. Source: Carpineti *et al.*, 2024

Climate treatment	Light treatment	Class 1 (#fruits/m²)	Class 2 (#fruits/m²)	Mildew (#fruits/m²)	Total production (#fruits/m²)	Not sellable (#fruits/m²)
Stable	Reference	591	104	190	885	122
Stable	High Blue	590	81	190	861	113
Fluctuation	Reference	542	90	225	857	125
Fluctuation	High Blue	527	80	189	797	126

Total number of harvested fruits per class. Source: Carpineti *et al.*, 2024

3. Peppers (*Capsicum spp.*)

Peppers, including bell peppers and hot varieties, are well-suited to vertical farming due to their compact growth and high yield. They require consistent temperatures and adequate light, which can be finely tuned in a controlled environment. Vertical farming allows for the cultivation of a wide range of pepper colours and flavours, catering to diverse consumer preferences.

4. Cucumbers (*Cucumis sativus*)

Cucumbers are commonly grown in vertical farming systems, where their climbing growth habit can be managed using trellises and support structures. They benefit from controlled humidity and temperature, which help prevent common issues like powdery mildew.

> While more challenging than leafy greens, these crops **can be profitable** if managed well. They often require more vertical space and may need manual pollination.

Medicinal Plants

1. Cannabis (*Cannabis Sativa*)

In regions where it is legally permitted, cannabis is a high-value crop often grown in vertical farming systems. The controlled environment allows for the precise regulation of factors such as light, humidity, and nutrient supply, which are crucial for optimizing cannabinoid profiles and yield. Vertical farming also provides security and discretion, which are important considerations for cannabis cultivation.

Scientific Report

Diniconazole Promotes the Yield of Female Hemp (*Cannabis sativa*) Inflorescence and Cannabinoids in a Vertical Farming System
Country: Republic of Korea
Publication Date: 30 May 2023
Main focus: This study investigates the effects of diniconazole on the growth and cannabinoid production in female hemp (*Cannabis sativa*) within a vertical farming system, aiming to optimize plant height and inflorescence yield.

Key findings:

- Treatment with 25 mg·L−1 diniconazole resulted in a 24.46% increase in inflorescence fresh weight and a 24.04% increase in inflorescence dry weight compared to the control group.
- The optimal concentration of 25 mg·L−1 diniconazole significantly enhanced the production of cannabinoids, with a 37.3% increase in CBD and a 27.6% increase in Δ9-THC in the inflorescences.
- Higher concentrations of diniconazole (50 mg·L−1 and above) led to a reduction in both plant growth and cannabinoid content.
- The study highlights the potential of using diniconazole in vertical farming to maximise the yield of high-value crops like female hemp, ensuring efficient space utilisation and enhanced profitability in controlled environments.

Reference: Hahm, S.; Lee, B.; Bok, G.; Kim, S.; Park, J. Diniconazole Promotes the Yield of Female Hemp (*Cannabis sativa*) Inflorescence and Cannabinoids in a Vertical Farming System. *Agronomy* **2023**, *13*, 1497. DOI: 10.3390/agronomy13061497

Pictures of female hemp (*Cannabis sativa* L. 'Hot blonde') on harvest day of diniconazole applications. The control group (**A**) did not receive the additional diniconazole treatment. Harvest day is 35 days after reproductive growth transition. (**A**) Control; (**B**) 25 mg·L−1; (**C**) 50 mg·L−1; (**D**) 100 mg·L−1; (**E**) 200 mg·L−1; and (**F**) 400 mg·L−1. Source: Hahm *et al.*, 2023

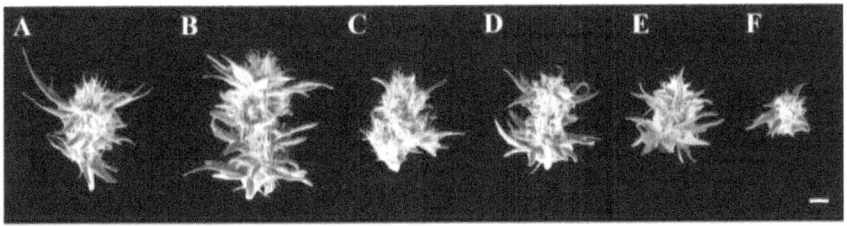

Pictures of female hemp (*Cannabis sativa* L. 'Hot blonde') apical inflorescences on harvest day of diniconazole applications. The control group (A) did not receive the additional diniconazole treatment. Harvest day is 35 days after reproductive growth transition. Scale bars 1 cm; (A) Control; (B) 25 mg·L−1; (C) 50 mg·L−1; (D) 100 mg·L−1; (E) 200 mg·L−1; and (F) 400 mg·L−1. Source: Hahm *et al.*, 2023

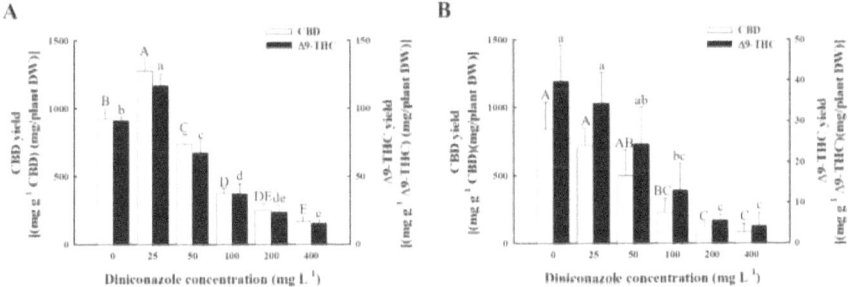

Total yield of major cannabinoids of female hemp (*Cannabis sativa* L. 'Hot blonde') inflorescence (**A**) and leaf (**B**) following diniconazole applications. The control group did not receive the additional diniconazole treatment. Harvest day is 35 days after reproductive growth transition. Data are means ± SD (n = 3). Different letters indicate significant differences among treatments at the level of 5%, according to Tukey's test. Source: Hahm *et al.*, 2023

2. Aloe Vera (*Aloe barbadensis miller*)

Aloe Vera is a popular medicinal plant grown in vertical farming due to its therapeutic properties, particularly for skin treatments. It requires well-drained soil and consistent temperatures, conditions that are easily maintained in a controlled environment. Vertical farms can produce Aloe Vera with high-quality gel, used in various cosmetic and medicinal products.

3. Echinacea (*Echinacea purpurea*)

Echinacea, often used to boost the immune system, is well-suited to vertical farming. It requires controlled lighting and temperature conditions, which can be efficiently provided in a vertical setup. This environment ensures the consistent production of high-quality Echinacea, which is commonly used in herbal supplements and teas.

4. Lavender (*Lavandula angustifolia*)

Lavender is cultivated in vertical farming for its calming properties and use in aromatherapy. It prefers a well-drained environment and specific light conditions, which can be easily managed in a controlled setting. Vertical farming enables the production of high-quality lavender, used in essential oils, skincare products, and as a natural remedy for relaxation and stress relief.

5. Chamomile (*Matricaria chamomilla*)

Chamomile, known for its soothing effects and use in herbal teas, can be effectively grown in vertical farms. It requires careful management of light and temperature to maintain its delicate flowers. Vertical farming systems allow for the year-round production of chamomile, ensuring a consistent supply for use in teas, tinctures, and topical applications.

Scientific Report

The Agro-Economic Feasibility of Growing the Medicinal Plant *Euphorbia peplus* in a Modified Vertical Hydroponic Shipping Container
Country: Belgium
Publication Date: 17 March 2022
Main focus: This study investigates the agronomic feasibility and economic viability of cultivating the medicinal plant *Euphorbia peplus* in a modified vertical hydroponic shipping container.

Key findings: The research identified that the fresh crop cost per kilogram ranged from €185 to €59, depending on the optimization of biomass yield and area surface. Ethyl acetate extraction at 120°C provided the best yield of 43.8 mg/kg at a cost of €38 per mg. Despite challenges in economic feasibility for pharmaceutical gel production, factors like improved yield and optimised conditions could potentially enhance profitability.

Reference: Bafort, F.; Kohnen, S.; Maron, E.; Bouhadada, A.; Ancion, N.; Crutzen, N.; Jijakli, M.H. The Agro-Economic Feasibility of Growing the Medicinal Plant *Euphorbia peplus* in a Modified Vertical Hydroponic Shipping Container. *Horticulturae* **2022**, *8*, 256. DOI: 10.3390/horticulturae8030256

Modified vertical hydroponic shipping container. Source: Bafort *et al.*, 2022

Cultivation area inside the modified vertical hydroponic shipping container. Thirty-six cultivation trays were placed on either side of the alley. Each cultivation tray could hold 24 plants distributed in bands of 8 plants in the cultivation tray as follows: along the wall, in the centre of the tray, and along the alley. Source: Bafort *et al.*, 2022

Mean and standard deviation of total ingenol (mg/kg) measured in aliquots of 24 plants in two independent hydroponic cultivations of *E. peplus* in three substrates (rockwool, coco fibre, and clay beads). Student's comparison of means was applied with 95% confidence; means that do not share a letter are significantly different. Source: Bafort *et al.*, 2022

	Trial 1		Trial 2	
Substrate	Number of Aliquots	Total Ingenol	Number of Aliquots	Total Ingenol
Clay beads	2	61.8 ± 1.57 a	6	60.0 ± 12 a
Coco Fiber	2	59.9 ± 3.84 a	6	63.5 ± 8.84 a
Rockwool	2	61.7 ± 1.41 a	6	61.8 ± 5.17 a

Total ingenol mean content and standard deviation in the aerial parts of *E. peplus* as a function of light intensity or plant localization in two independent experiments. Total ingenol was measured in aliquots of 8 plants. Student's comparison of means was applied with 95% confidence; means that do not share a letter are significantly different. Light intensity: Trial 1: PAR 250: N = 18 aliquots; PAR 500: N = 15 aliquots; trial 2: N = 18 aliquots. Localization: Trial 1: N = 11 aliquots; trial 2: N= 12 aliquots. Source: Bafort *et al.*, 2022

		Trial 1			Trial 2		
Light Intensity	($\mu mol\, m^{-2}\, s^{-1}$)	250	500		250	500	
Total Ingenol	(mg/kg)	74.7 ± 25.4 a	65.7 ± 19.6 a		60.9 ± 2.61 a	62.4 ± 3.08 a	
Localization		Center	Alley	Wall	Center	Alley	Wall
Total Ingenol (mg/kg)		71.5 ± 24.6 a	72.4 ± 24.1 a	68.0 ± 22.3 a	62.8 ± 1.81 a	61.3 ± 3.63 a	60.9 ± 2.88 a

Cultivation costs of *Euphorbia peplus* and Romaine lettuce in several scenarios according to the light intensity and growing surface, generating fresh biomass, output, capex, and opex. The cost per kg of fresh biomass includes capex and opex; the contribution of each particular cost to the total cost was calculated as a percentage. The values mentioned under the "R&D" container are experimental results, and the values mentioned under the "R&D+" container and "Commercial container" are projected results. Source: Bafort et al., 2022

Crop		*Euphorbia peplus*							Romaine Lettuce	
Light	($\mu mol\,m^{-2}\,s^{-1}$)	150				500			150	
Fresh Biomass per crop	(g)	32.7				102			102	
		R&D container		R&D+ container		R&D container		R&D+ container		Commercial container
Total Growing surface (sqm)	sqm	30		40		30		40		50
Fresh Biomass (incl. 5% quality loss)	(kg/yr/sqm)	6.4		6.34		19.97		19.39		34.04
OUTPUT	(kg/yr)	192		254		599		776		1702
CAPEX	(EUR/sqm)	3500		3000		3833.33		3375		3100
CAPEX (15-yr depreciation)	(EUR/yr)	7000		8000		7667		9000		10,333
OPEX										
Technical Staff at €210/day	(EUR/yr)	12,023		13,385		12,023		13,385		19,467
Engineer staff at €310/day	(EUR/yr)	4437		4394		4437		4394		1465
Director staff at €600/day	(EUR/yr)	2147		2126		2147		2126		709
Electricity at €0.2/kW	(EUR/yr)	6650		8845		9177		12,206		10,964
Water at €4.84/m³	(EUR/yr)	33		45		34		45		50
Seeds	(EUR/yr)	0		0		0		0		75
Fertilizer	(EUR/yr)	1001		1319		1001		1319		810
Substrates (rockwool)	(EUR/yr)	1255		1057		1255		1057		934
pH adjustors	(EUR/yr)	47		62		47		62		76
Container maintenance	(EUR/yr)	1001		1301		1001		1301		1620
TOTAL	(EUR/yr)	28,597		33,144		31,124		36,505		36,176
COST per kg of fresh biomass										
CAPEX (15-yr depreciation)	(EUR/kg)	36.44	20%	31.54	19%	12.80	20%	11.60	20%	6.07	22%
Labor Technical staff	(EUR/kg)	62.59	34%	52.81	34%	20.07	31%	17.27	29%	11.44	42%
Labor Eng. staff	(EUR/kg)	23.10	12%	17.32	11%	7.41	11%	5.87	10%	0.86	3%
Labor Director staff	(EUR/kg)	11.18	6%	8.38	5%	3.58	6%	2.74	5%	0.42	2%
Electricity	(EUR/kg)	34.62	19%	34.87	21%	15.32	24%	15.74	27%	6.44	24%
Water	(EUR/kg)	0.18	0%	0.18	0%	0.06	0%	0.06	0%	0.03	0%
Seeds	(EUR/kg)		0%		0%		0%		0%	0.04	0%
Fertilizer	(EUR/kg)	5.21	3%	5.20	3%	1.67	3%	1.70	3%	0.48	2%
Substrates (rockwool)	(EUR/kg)	6.53	4%	6.53	4%	2.09	3%	2.14	4%	0.55	2%
pH adjustors	(EUR/kg)	0.24	0%	0.24	0%	0.08	0%	0.08	0%	0.04	0%
Container maintenance	(EUR/kg)	5.21	3%	5.13	3%	1.67	3%	1.68	3%	0.96	3%
TOTAL	(EUR/kg)	185.31		162.22		64.74		58.67		27.33	

Edible Flowers

Edible flowers provide diversity and can attract higher prices in specialised markets, such as high-end restaurants.

1. Nasturtium (*Tropaeolum majus*)

Nasturtiums are popular edible flowers grown in vertical farming for their vibrant colours and peppery flavour. They thrive in controlled environments where temperature and light can be optimised for consistent flowering. Vertical farms can produce nasturtiums year-round, offering fresh, pesticide-free blossoms for use in salads, garnishes, and as a decorative element in dishes.

Scientific Report

Continuous Lighting and High Daily Light Integral Enhance Yield and Quality of Mass-Produced Nasturtium (*Tropaeolum majus* L.) in Plant Factories
Country: Japan
Publication Date: 12 June 2021
Main focus: This study investigates the effects of continuous lighting (CL) and different daily light integrals (DLIs) on the growth, secondary metabolites, and light use efficiency (LUE) of nasturtium in plant factories with artificial lighting (PFALs).

Key findings:
- Continuous lighting significantly increased leaf production, secondary metabolites, and LUE compared to non-continuous lighting. Under CL, increasing the DLI from 17.3 to 34.6 mol m^{-2} d^{-1} enhanced the biomass without causing physiological stress. Specific numerical insights include:
- Leaf Fresh Weight (FW): Increased by 30-40% under 24-hour lighting compared to 16-hour lighting at early growth stages.
- Light Use Efficiency (LUE): Higher under continuous lighting, with significant increases noted at various stages of plant growth.
- Daily Light Integral (DLI) Impact: Yield and biomass showed positive linear relationships with increasing DLI levels under continuous lighting.

Reference: Xu, W.; Lu, N.; Kikuchi, M.; Takagaki, M. Continuous Lighting and High Daily Light Integral Enhance Yield and Quality of Mass-Produced Nasturtium (*Tropaeolum majus* L.) in Plant Factories. *Plants* **2021**, *10*, 1203. DOI: 10.3390/plants10061203

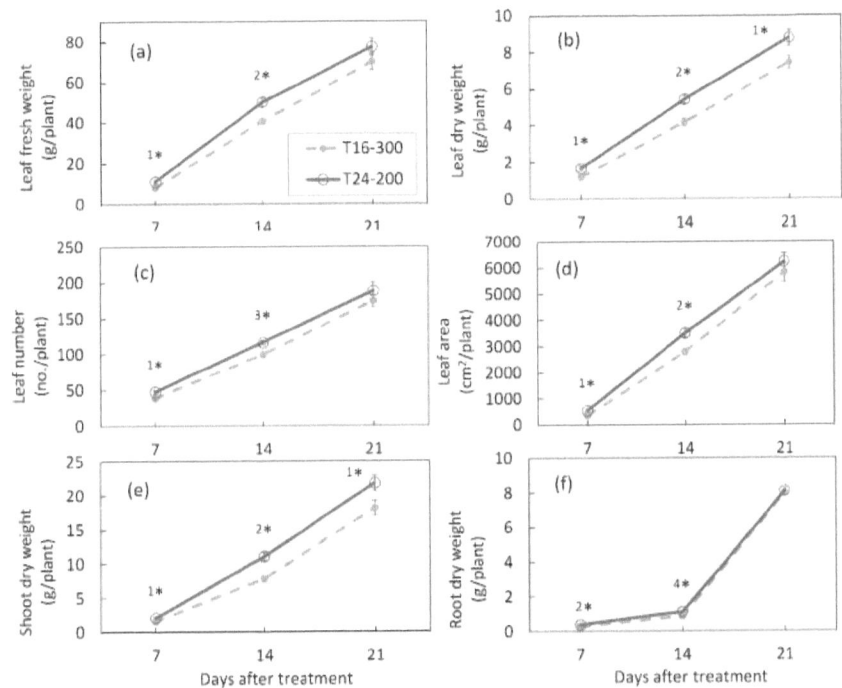

Leaf fresh (a) and dry weight (b), leaf numbers (c), leaf area (d), shoot dry weight (e), and root dry weight (f) of nasturtium grown under T16-300 and T24-200 at 7, 14, and 21 days after treatment. Values are the means ± SE (n = 6). Asterisks indicate significant differences between the treatments (1*, $p < 0.05$; 2*, $p < 0.01$; 3*, $p < 0.001$, and 4*, $p < 0.0001$), determined by the t-test. Plant density during the 1st, 2nd, and 3rd weeks was 21.5, 16.7, and 11.1 plants m−2, respectively. Source: Xu W et al., 2021

2. Pansy (*Viola tricolor var. hortensis*)

Pansies are widely cultivated in vertical farming systems for their colourful petals and mild, slightly sweet flavour. They require cool temperatures and moderate light, which can be easily maintained in a controlled setting. Vertical farms can grow pansies efficiently, ensuring a steady supply of fresh, edible flowers for culinary uses and decorative purposes, particularly in desserts and salads.

3. Marigold (*Tagetes spp.*)

Marigolds, known for their bright orange and yellow blooms, are grown in vertical farming for their edible petals with a slightly citrusy flavour. They flourish in warm, well-lit environments, which are easily provided in vertical farming setups. The controlled conditions help ensure the consistent quality and availability of marigolds, which are used to add colour and a unique flavour to dishes, teas, and garnishes.

Other Crops

1. Mushrooms (*Agaricus bisporus*)

Mushrooms, such as shiitake, oyster, and button varieties, are well-suited for vertical farming due to their ability to thrive in controlled environments with minimal light, making efficient use of vertical space while providing a high-value crop with relatively quick growth cycles.

Scientific Report

Effects of Different Lighting Conditions on Growth, Yield and Nutrient Content of White Oyster Mushroom in Vertical Farm

Country: Bangladesh
Publication Date: 15 December 2021
Main focus: This research examines the impact of various lighting conditions on the growth, yield, and nutrient content of white oyster mushrooms cultivated in vertical farms.

Key findings:
- The study found significant differences in growth and yield under different lighting conditions and structural layers.
- In the dark condition, the average growth and yield were highest, with a stalk height of 3.87 cm, cap size of 11.89 cm, stipe size of 2.65 cm, 15.78 crops per bag, a fresh weight of 60.03 g, and a dry weight of 17.03 g.
- Conversely, the lowest growth and yield were observed in sunlight, with a stalk height of 1.59 cm, cap size of 3.67 cm, stipe size of 1.51 cm, 4.22 crops per bag, a fresh weight of 18.38 g, and a dry weight of 4.47 g.
- Nutrient content was also highest in the dark condition, suggesting that maintaining absolute darkness is crucial for high-quality white oyster mushroom production.

Reference: Al Mamun, M. R., Deb, I., Hridoy, T., Soeb, M. J. A., & Shammi, S. Effects of Different Lighting Conditions on Growth, Yield and Nutrient Content of White Oyster Mushroom in Vertical Farm. *European Journal of Agriculture and Food Sciences* **2021**, 3(6), 61–67. DOI: 10.24018/ejfood.2021.3.6.418

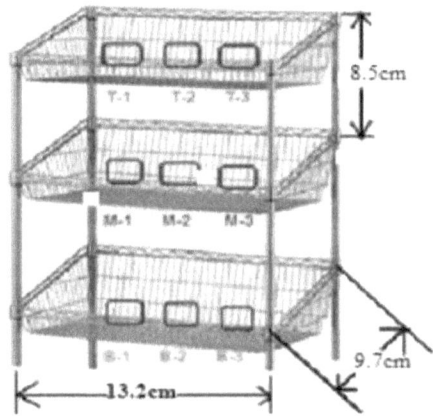

Design of the vertical farm for white oyster mushrooms (3D). Source: Al Mamun et al., 2021

Observation No.	Temperature		Humidity (%)	Irrigation Water (ml)
	Room (°C)	Outside (°C)		
1	24.3	24.8	67	
2	25.2	28.3	71	
3	25.8	27.5	69	
4	23.6	25.2	70	
5	22.8	24	68	
6	23	25.5	70	
7	23.8	25.5	72	
8	26	28.3	60	
9	27.2	28.5	65	1200
10	25.3	26.8	86	
11	23.6	24	66	
12	26	26.6	69	
13	25.6	27	71	
14	22	23.2	74	
15	22.8	24.6	72	
16	18	19.8	59	
17	27	28.2	63	
18	27.9	29.3	61	

Daily weather conditions of the vertical farm for white oyster mushrooms.
Source: Al Mamun et al., 2021

2. Soybean (*Glycine max*)

There are present challenges in vertical farming due to their height and light requirements, but ongoing research into dwarf varieties and optimised lighting systems shows potential for future integration, offering a protein-rich crop option for vertical farms. At the same time, with current electricity prices, it is unlikely to justify production of simple protein crops in vertical farming or promote it as a solution to meet global protein needs [Righini *et al.*, 2024]

Scientific Report

Protein plant factories: production and resource use efficiency of soybean proteins in vertical farming
Country: Netherlands
Publication Date: 25 March 2024
Main focus: This study investigates the production and resource use efficiency of soybean proteins in vertical farming (VF) systems compared to traditional open-field cultivation, with a focus on controlled environment agriculture (CEA) for efficient macronutrient production.

Key findings:
- The protein yield per square metre of crop in VF was about eight times higher than in the open field.
- To meet the total yearly protein requirement for a reference adult solely through VF, 20 m² of crop area, 2.4 m³ of water, and 16 MWh of electricity are required, versus 164 m², 111 m³, and 0.009 MWh in the field.
- Obelix cultivar yielded 18% more than Viola. Protein concentration was on average 8% higher in Viola compared to Obelix.

Reference: Righini, I., Graamans, L., van Hoogdalem, M., Carpineti, C., Hageraats, S., van Munnen, D., Elings, A., de Jong, R., Wang, S., Meinen, E., Stanghellini, C., Hemming, S. and Marcelis, L.F., Protein plant factories: production and resource use efficiency of soybean proteins in vertical farming. *J Sci Food Agric* **2024**, 104: 6252-6261. DOI: 10.1002/jsfa.13458

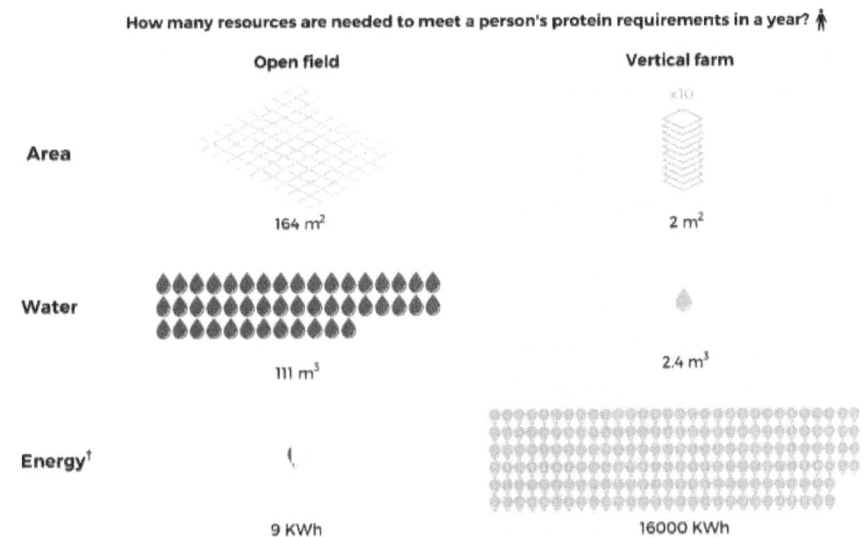

Land area, water and energy required in the open-field and VF cultivation to meet a person's protein requirements in a year (0.83 g protein kg−1 bodyweight, 70 kg reference adult). In the figure, the area usage for VF is 10 times more efficient compared to the value of 20 m2 reported in the text, as it assumes 10 cultivation tiers as an example.
Source: Righini et al., 2024

3. Wheat (*Triticum aestivum*)

Cultivation in vertical farms is **currently limited by its height and extensive root system**, but advancements in hydroponic systems and the development of shorter, high-yielding varieties may make it a viable option in the future, potentially changing urban grain production.

Scientific Report

Wheat Yield Potential in Controlled-Environment Vertical Farms

Country: United States
Publication Date: 27 July 2020
Main focus: The study examines the potential of vertical farming for wheat production, exploring the feasibility of high-yield indoor wheat cultivation under controlled environmental conditions.

Key findings: The research demonstrated that wheat grown in a 10-layer indoor vertical farm could achieve yields between 700 ± 40 t/ha and a maximum of 1,940 ± 230 t/ha annually, which is 220 to 600 times the current world average wheat yield of 3.2 t/ha. However, the high energy costs, particularly for artificial lighting, present significant economic challenges.

Reference: Asseng, S., Guarin, J. R., Raman, M., & Gauthier, P. P. G. Wheat yield potential in controlled-environment vertical farms. *Proceedings of the National Academy of Sciences* **2020**, 117(32), 19131-19135. DOI: 10.1073/pnas.2002655117

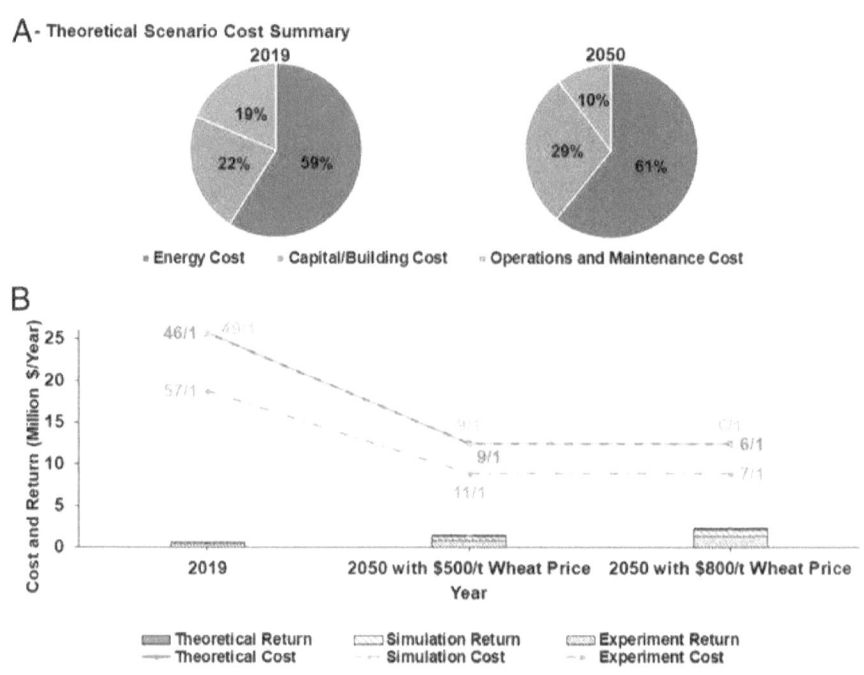

Annual cost and return for indoor wheat farming.

(A) Pie charts showing 2019 (Left) and 2050 (Right) breakdown of costs as percentages for a 1-ha, 10-layer indoor wheat growing scenario with an adapted high harvest index cultivar and capital and building costs financed at 5% per year. A breakdown of the costs for simulation and experiment scenarios are provided in the full text of the research paper.
(B) Total annual cost of wheat production (lines) and annual returns (stacked bars) for a 1-ha, 10-layer facility for theoretical (red), simulation (green), and experimental (blue) indoor wheat growing scenarios, assuming wheat prices of $200/t in 2019 (2) and $500/t and $800/t in 2050 (based on a likely increase in the future price and premium price for pesticide-free production). The 2050 cost is the same for the $500/t and $800/t wheat price scenarios. Data point labels are cost/return ratios for each scenario. Error bars show SEM when larger than symbols.
Source: Asseng *et al.*, 2020

4. Ornamental Plants

Ornamental plants are suitable for vertical farming because they **require less space**, can **thrive in controlled environments**, and offer **high aesthetic and economic value**

Scientific Report

Soilless Propagation of *Haberlea rhodopensis* Friv. Using Different Hydroponic Systems and Substrata
Country: Bulgaria
Publication Date: June 2020
Main focus: This study explores the soilless propagation of the endemic plant *Haberlea rhodopensis* using various hydroponic systems and substrates, aiming to improve rooting and rosette formation for conservation and ornamental purposes.

Key findings:

- Leaf rooting and survival rates were up to 86.7%, with the best results obtained using IBA-treated leaves in perlite/agrolava substrate on the vertical system, achieving 46.7% leaves with rosettes and 2.9 rosettes per leaf.
- Rosettes acclimated to soil reached up to 10 cm in diameter within a year.

Reference: Traykova B.D., Stanilova M.I., Soilless Propagation of *Haberlea rhodopensis* Friv. Using Different Hydroponic Systems and Substrata, Ecologia Balkanica **2020**, Vol. 12, Issue 1, pp. 111-121.

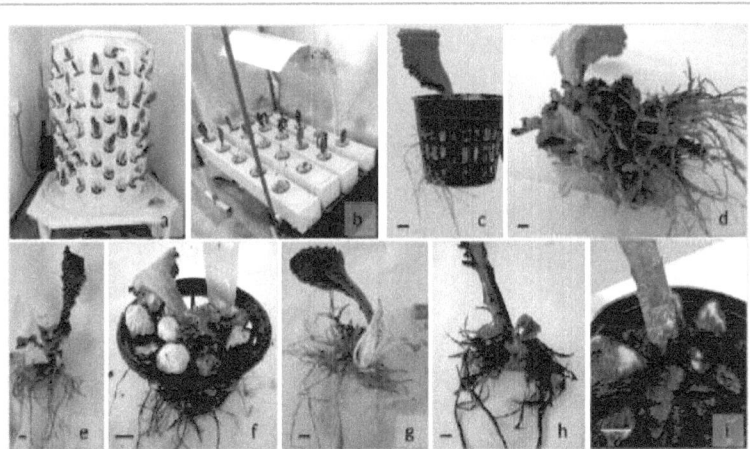

H. rhodopensis rosettes formation on aero-hydroponic systems.
a) Vertical Green Diamond system; b) Horizontal Aeroflo-20 system; c) Roots on variant

P-AP-IBA, 12 weeks old; d) Numerous rosettes formed at the leaf base in variant MW-AP-IBA; e) A single rosette formed in variant MW-AP-C; f) Rosettes formed on Aeroflo-20 in variant AP-6m-L; g) Etiolated rosette formed under the agrolava pebbles; h) Partial putrefaction of the root system; i) Algae development on the substrata. Scale bars = 10 mm. Souce: Traykova & Stanilova, 2020

Crop selection is always an **ongoing process of optimization**, with farms continually evaluating new varieties and adjusting their crop mix based on market trends, production efficiencies, and technological advancements.

Crop Pests and Diseases

Considering indoor vertical farming, crop pests and diseases pose **significant challenges in vertical farming systems**, despite the controlled environment. While these systems offer some protection against traditional outdoor threats, they are **not immune to biological adversaries.**

In vertical farms, the enclosed growing environment has both advantages and disadvantages regarding pests and diseases. On the positive side, it **limits exposure to many outdoor pests and pathogens**. However, if an infestation or infection does occur, it can **spread rapidly due to the close proximity of plants and the favourable growth conditions** that benefit both crops and their potential adversaries.

Common pests in vertical farming systems include aphids, whiteflies, spider mites, and thrips, which can quickly multiply, feeding on plant sap and potentially transmitting viruses. **Fungal diseases** such as powdery mildew and botrytis can also thrive in the humid conditions often found in vertical farms.

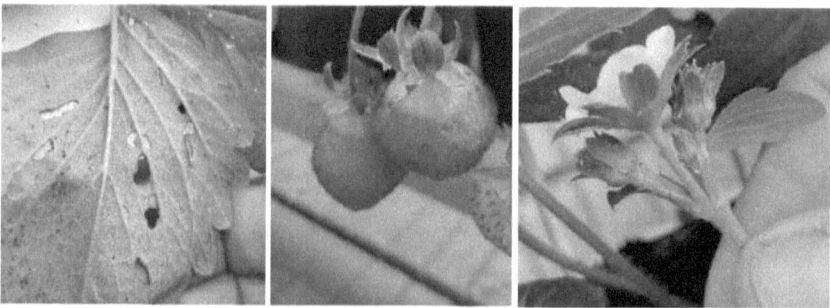

Caterpillar of leaf wasp (left) mildew on fruit (central) aphids on flower truss (right) on DAT 19.
Source: Carpineti *et al.*, 2024

Although less common, **bacterial and viral infections** can be devastating if introduced to the system.

Prevention involves **strict hygiene protocols**, such as clean room practices for workers entering growing areas, regular sanitization of equipment, and careful monitoring of plant material brought into the facility. **Advanced air filtration systems** are often employed to reduce the risk of airborne pathogens and pests entering the environment.

Integrated Pest Management (IPM) strategies are particularly effective in vertical farming. These approaches combine biological, cultural, physical, and chemical tools to manage pests and diseases while minimising environmental impact.

For example, biological control methods, such as **introducing beneficial insects** like ladybugs or predatory mites, can be highly effective in the controlled environment of a vertical farm.

Scientific Report

Open Vertical Farms: A Plausible System in Increasing Tomato Yield and Encouraging Natural Suppression of Whiteflies
Country: Slovenia
Publication Date: 2022
Main focus: The study evaluates the effectiveness of open vertical farming in increasing tomato yields and recruiting ecological service providers for the natural suppression of whiteflies.

Key findings:
- The open vertical farm produced significantly higher tomato yields compared to the traditional horizontal farm.
- The vertical farm design, with three arrays at different heights, utilized only 1.8 m² of land space, demonstrating efficient land use.
- The mean number of predatory spiders in the vertical farm from 6 to 10 weeks after transplanting successfully suppressed whitefly populations compared to the horizontal farm.
- The study suggests that outdoor vertical farming could be a viable solution to land shortages and a sustainable system for integrated pest management.

Reference: Mustapha, S., Musa, A. K., Apalowo, O. A., Lawal, A. A., Olayiwola, O. I., Bamidele, H. O., & Uddin II, R. O. Open vertical farms: a plausible system in increasing tomato yield and encouraging natural suppression of whiteflies. Acta Agriculturae Slovenica **2022**, 118(2), 1–9. DOI: 10.14720/aas.2022.118.2.2272

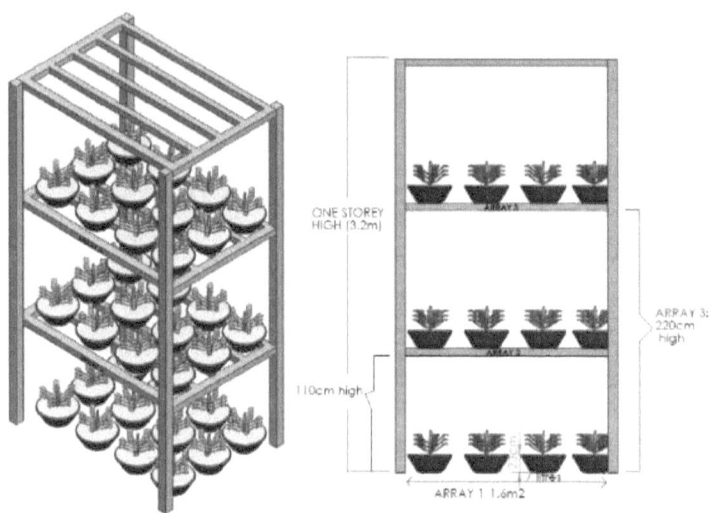

Schematic diagram of the open-air vertical structure used for the experiment to increase tomato yield and encourage natural suppression of whiteflies. Source: Mustapha *et al.*, 2022

Showing presence of spiders in the vertical farm. Pictures a, b: green spiders observed using camouflage to hunt for prey; pictures c, d: two different species of spider on their web to capture prey. Source: Mustapha et al., 2022

Number of *Bemisia tabaci* on the vertical and horizontal farm.
Source: Mustapha et al., 2022

Farm Type		Population of *B. tabaci* in horizontal and vertical farms (WAT)									
		1	2	3	4	5	6	7	8	9	10
Horizontal	Raised bed	9.15^d	8.61^c	7.26^b	11.75^c	13.22^d	9.54^d	8.28^b	9.51^c	13.35^c	8.84^b
	Flat bed	12.81^c	9.75^d	10.64^d	10.51^c	14.69^c	11.84^c	9.12^c	7.04^b	9.38^b	10.80^c
Vertical	Array 1	8.72^c	7.75^b	13.79^c	6.54^a	7.64^a	4.99^b	0.00^a	0.00^a	0.00^a	0.00^a
	Array 2	6.52^a	7.68^a	6.96^a	8.87^d	7.38^a	3.73^b	0.00^a	0.00^a	0.00^a	0.00^a
	Array 3	5.30^a	11.00^c	8.08^c	6.33^b	5.80^a	0.00^a	0.00^a	0.00^a	0.00^a	0.00^a
SEM		0.82	0.50	1.44	1.35	1.24	1.05	1.28	0.98	0.60	0.61

Superscripts within column indicates mean rank number according to Kruskal-Wallis Test, with a = rank 1, b = rank 2, c = rank 3, d = rank 4 and e = rank 5; 1 being the lowest to 5 the highest rank, SEM = Standard error of mean

Number of predatory spiders on the vertical and horizontal farm. Source: Mustapha et al., 2022

Farm Type		Population of predatory spiders in horizontal and vertical farms (WAT)							
		3	4	5	6	7	8	9	10
Horizontal	Raised bed	0.00[a]	0.33[a]	0.00[a]	0.67[a]	1.00[b]	1.00[b]	6.30[b]	0.00[a]
	Flat bed	0.00[a]	0.33[a]	0.33[b]	0.67[b]	1.70[a]	0.30[a]	4.30[a]	0.00[a]
Vertical	Array 1	0.00[a]	0.33[a]	9.33[c]	5.00[c]	15.00[c]	21.00[c]	23.70[c]	16.30[b]
	Array 2	0.67[c]	1.00[b]	7.67[c]	10.33[d]	20.30[d]	22.30[d]	21.70[d]	20.00[c]
	Array 3	0.33[b]	3.33[c]	8.67[d]	13.00[e]	34.00[e]	27.00[e]	21.00[c]	35.00[d]
SEM		0.19	0.73	1.25	1.10	4.02	2.49	3.61	2.31

Superscripts within column indicates mean rank number according to Kruskal-Wallis Test, with a = rank 1, b = rank 2, c = rank 3, d = rank 4 and e = rank 5; 1 being the lowest to 5 the highest rank, SEM = Standard error of mean

Correlation between no. of B. tabaci, no. of spiders and fruit yield in the vertical and horizontal farm. Source: Mustapha et al., 2022

		Horizontal farm		Vertical Farm	
		BTH	SPH	BTV	SPV
Horizontal farm	BTH	-	-	-	-
	SPH	-0.318		-0.788	
	FYH	-0.28	-0.256	0.693	-0.699
Vertical farm	BTV	0.363			
	SPV	-0.333	0.802	-0.999**	
	FYV	-0.083	0.303	-0.813	0.806

*Correlation is significant at the 0.05 level (2-tailed), ** Correlation is significant at the 0.01 level (2-tailed)
BTH= B. tabaci horizontal farm, BTV= B. tabaci vertical farm, SPH= Spiders horizontal farm, SPV= Spiders vertical farm, FYH= Fruit yield horizontal farm, FYV= Fruit yield vertical farm, - = no correlation

Regular monitoring using advanced sensing technologies and AI-driven systems allows for early detection of issues, enabling rapid response and targeted treatment before problems become visible.

When interventions are necessary, there is a range of options. **Organic pesticides** and **fungicides** can be used carefully to avoid disrupting beneficial organisms or leaving harmful residues on crops. **Physical controls**, such as sticky traps for flying insects or **careful pruning of affected plant parts**, can also be effective.

Water-borne pathogens, a unique challenge in closed-loop vertical farming systems, can be managed with advanced water treatment technologies like **UV sterilisation and ozonation**.

Outdoor vertical farming gardens face unique challenges in pest and disease management. These systems are **exposed to a wider range of environmental factors and potential threats**, such as local insect populations and wind-borne spores. Outdoor vertical gardens must contend with **seasonal pest cycles** and may require more robust IPM strategies, including physical barriers, pest-repelling companion plants, and the careful timing of plantings to avoid peak pest seasons.

Scientific Report

Vertical Farming Systems Bring New Considerations for Pest and Disease Management
Country: United Kingdom
Publication Date: 25 February 2020
Main focus: This study explores the unique challenges and considerations for pest and disease management in vertical farming systems, compared to conventional protected horticulture.

Key findings:
- The study highlights that vertical farming introduces specific pest and disease management issues due to the stacked arrangement of crops and controlled environment settings.
- Vertical farms face challenges such as vertical and horizontal pest movement, the need for tailored management approaches for different growing levels, and the effects of environmental gradients on pest and pathogen behaviour.
- The study emphasises the importance of integrating pest management strategies, including biological controls and semiochemical deployment, to address these challenges effectively.
- The research also points out that despite the controlled environment, vertical farms cannot completely eliminate pest and disease risks, necessitating robust control measures and system design optimizations.

Reference: Roberts, J. Vertical Farming Systems Bring New Considerations for Pest and Disease Management. *Horticultural Science Journal*, **2020**, Issue 2. DOI: 10.1016/j.hortsci.2020.02.001.

Tipburn is not a disease but one of the **physiological disorders affecting plants**, particularly leafy vegetables like lettuce.

It is characterised by **necrosis (death) of the leaf margins**, often caused by a deficiency of calcium in the plant tissues, which can result from environmental conditions or improper nutrient management.

Factors like **high humidity, rapid growth, and inadequate transpiration** can increase calcium deficiency, leading to tipburn.

Scientific Report

Mitigating Tipburn Through Foliar Calcium Application in Indoor Hydroponically Grown Mini Cos Lettuce
Country: Turkey
Publication Date: 8 May 2024
Main focus: This study evaluates the effectiveness of different calcium dosages applied via foliar spray to mitigate tipburn in hydroponically grown green and red mini Cos lettuce varieties, Thespian and Suntred.

Key findings:
- Foliar application of 800 ppm CaO resulted in the highest plant height (23.8 cm) and leaf fresh weight (202.3 g) in Suntred lettuce.
- In Thespian lettuce, the 1000 ppm CaO treatment yielded the highest leaf fresh weight at 157.0 g and the highest calcium content at 3.62%.
- Calcium application significantly increased the plant weight and mitigated tipburn symptoms, with Suntred showing more promising results compared to Thespian.

Reference: Ikiz B, Dasgan HY, Oz BC. Mitigating Tipburn Through Foliar Calcium Application in Indoor Hydroponically Grown Mini Cos Lettuce. *BIO Web Conf.* **2024**. 85:01003. DOI: DOI: 10.1051/bioconf/20248501003

Thespian and Suntred lettuce varieties grown in water culture in the climate chamber.
Source: Ikiz *et al.*, 2024

Effects of CaO treatments at different ppm on plant height, plant width, plant circumference, and stem diameter of Thespian lettuce cultivar.
Source: Ikiz et al., 2024

Apps	Plant Height (cm)	Plant Width (cm)	Plant Circumference (cm)	Body Diameter (cm)
1	22,51	32,3	38,0	14,6
2	23,51	31,0	41,0	14,4
3	24,19	33,58	40,6	13,6
4	24,07	34,52	38,3	13,7
5	N.S.	N.S.	N.S.	N.S.

1: Control; 2: 200 ppm CaO; 3: 400 ppm CaO; 4: 800 ppm CaO; 5: LSD

Number of leaves, root fresh weight, root length, Leaf fresh weight of Thespian lettuce cultivar. Source: Ikiz et al., 2024

Apps	Number of Leaves (pcs/plant)	Root Fresh Weight (g)	Root Lenght (cm)	Leaf Fresh Weight (g/plant)
1	35,3	17,7	36,5 a	157,0 a
2	37,4	18,3	31,8 ab	133,1 ab
3	36,7	17,6	33,8 ab	148,5 ab
4	35,3	15	28,4 b	124,8 b
	N.S	N.S.	6,9	24,644

Plant height, plant width, plant circumference, and stem diameter values of Suntred red mini lettuce cultivar. Source: Ikiz et al., 2024

Appns	Plant Height (cm)	Plant Width (cm)	Plant Surroundings (cm)	Body Diameter (mm)
1	18,9 b	30,1 b	36,1 c	12,4 b
2	19,6 b	30,8 b	37.7 bc	13.1 ab
3	20,7 b	31,8 b	40,3 b	14,2 a
4	23,8 a	35,4 a	44,2 a	14,5 a
5	1,822	3,098	3,504	1,693

Number of leaves, root fresh weight, root length, and leaf fresh weight values of Suntred red mini lettuce cultivar. Source: Ikiz et al., 2024

Apps.	Number of leaves (pcs/plant)	Root Fresh Weight (g)	Root Length (cm)	Leaf Fresh Weight (g/plant)
1	29,7 b	14,9 b	32,6	128,8 a
2	29,7 b	17,8 b	33,7	143,0 b
3	31,1 b	18.2 ab	32,6	148,8 b
4	38,2 a	22,3 a	30,6	202,3 a
5	5,578	4,161	S.D.	30,812

Environment

The **environment in vertical farming** includes all aspects of growing conditions, including climate control, which manages temperature, humidity, and air circulation. It also involves carbon dioxide augmentation and the irrigation system. The environment and nutrient solution control unit plays a crucial role, serving as the headquarters for overall environment management.

Lighting and Energy Efficiency

Lighting in vertical farms typically relies on artificial sources, predominantly **LED (Light Emitting Diode) technology**, a type of energy-efficient light source that emits light when an electrical current passes through it.

The first practical visible-spectrum LED was created in 1962 by Nick Holonyak Jr., while working at General Electric in the United States. Today LEDs are favoured in vertical farming due to their **energy efficiency, low heat output, and precise spectral control**.

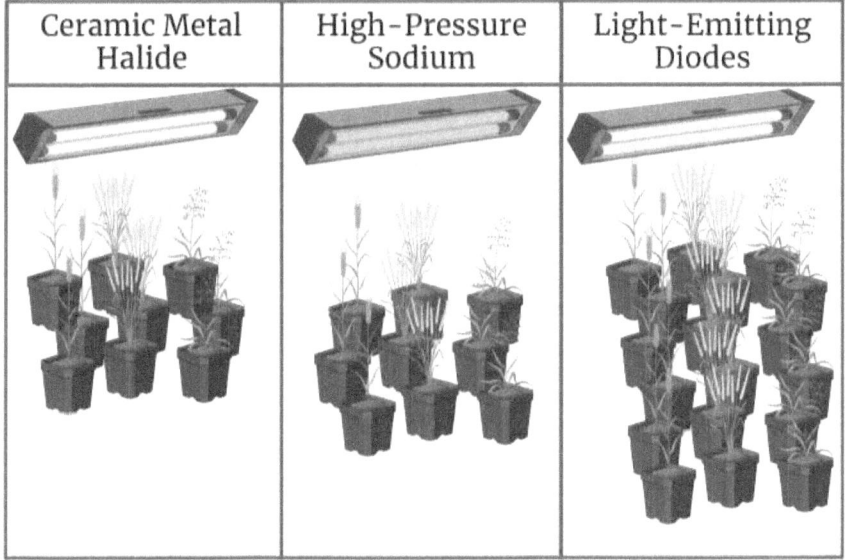

Effects comparison of supplementary lighting types on generation turnover. Aside from their overall energy efficiency and operating costs, light-emitting diodes (LEDs) outperform ceramic metal halide and high-pressure sodium lights in generations per year of barley, wheat, and oat crops. Source: Williams et al., 2024

Relevant parameters related to lighting in vertical farming systems.
Source: Najero *et al.*, 2023

Parameter	Description
Electromagnetic spectrum	The wavelength of light determines the distribution of energy, including both visible and non-visible wavelengths. Lamps emit light across a range of electromagnetic radiation detectable by the human eye, from 380 nm to 780 nm, which falls between the ultraviolet and infrared parts of the spectrum and includes components of solar radiation.
Photosyntenthically active radiation (PAR)	Light used for photosynthesis consists of tiny particles or wavelets called photons. This light falls within the range of 400 to 700 nanometers (nm) in the radiation spectrum, which is the part that plants use for photosynthesis.
Photosynthetic Photon Flux Density (PPFD)	Light intensity on a surface is measured as photosynthetic photon flux density (PPFD) with units of $\mu mol \cdot m^{-2} \cdot s^{-1}$. PPFD includes about half the energy of sunlight and matches well with how quickly photosynthesis responds to changes in light intensity.
Photoperiod	All living things need both light and darkness to keep their internal processes in balance. This daily cycle of 24 hours comes from the Earth's rotation, causing changes in sunlight depending on the location, the time of year, and the time of day.
Daily Light Integral (DLI)	DLI measures the total amount of light a plant gets in one day. It's important for vertical farming because it helps farmers understand how much light their crops need to grow well.
Radiation use efficiency (RUE) and Light use efficiency (LUE)	RUE measures how effectively a crop uses absorbed light energy for biomass production, calculated as the ratio of biomass gained to photosynthetically active radiation (PAR) absorbed. In contrast, LUE (Light Use Efficiency) refers to the net CO_2 assimilation efficiency, indicating how well plants convert absorbed light into CO_2 absorption. LUE can be measured over short periods or daily at the leaf level.

Plants primarily **use red and blue light** for photosynthesis, so many vertical farm lighting systems focus on these wavelengths.

However, other lights, such as **green and white**, can enhance growth, improve plant morphology, and increase the nutritional quality of crops by providing a more balanced light spectrum.

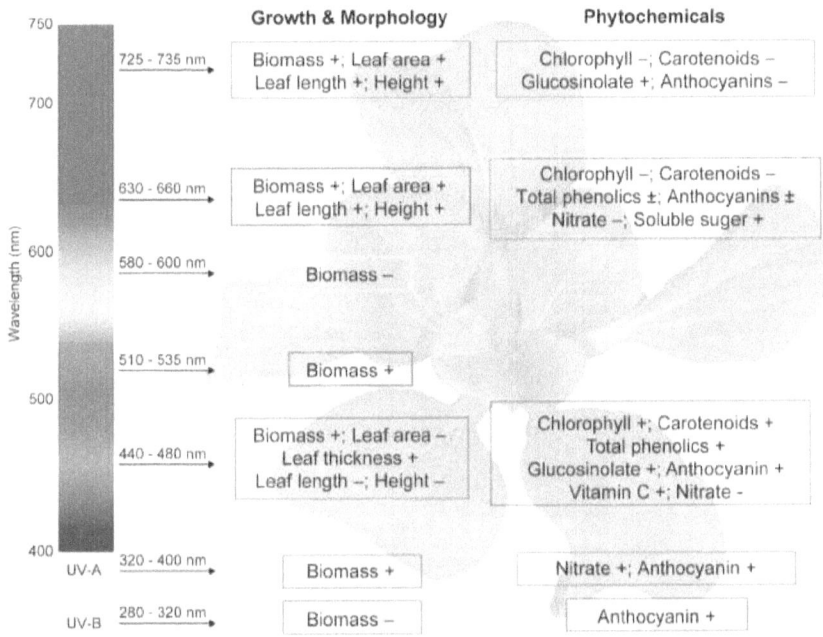

Responses of leafy greens to different light quality. The effects of an increase in the specific wavelength ranges on growth and phytochemical contents in leafy greens are summarised. These are based on the general trend observed among specific studies. A positive effect is depicted by the "+" sign while a negative impact is represented by the "-" sign. A "±" sign denotes varying results. Source: Wong et al., 2020

Red light is important for **flowering and fruit production**, while **blue light promotes vegetative growth**. By adjusting the balance of red and blue light, vertical farms can optimise plant growth for specific stages and desired outcomes.

However, **full-spectrum lights that mimic natural sunlight** are also employed to promote more natural plant development and improve other aspects of plant growth, such as flowering and fruiting.

Full-Spectrum Lights provide a more balanced range of wavelengths, closely replicating natural sunlight. They can enhance the overall health and development of plants, supporting processes beyond photosynthesis.

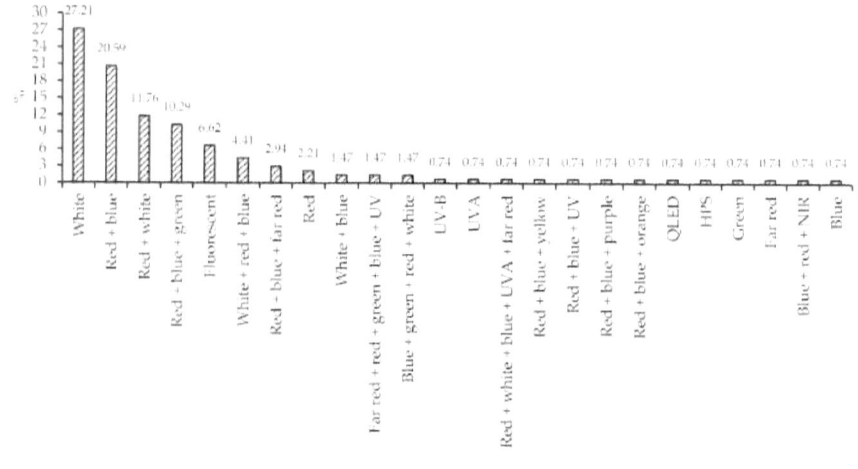

Illumination spectra used in vertical farming research for the growth and development of horticultural, ornamental, fruit, and aromatic plants. n = 136. Source: Najaro et al., 2023

Impacts of LED spectral quality, intensity and photoperiod on leafy vegetables grown indoors and in the tropical greenhouse in Singapore. Source: He, 2024

Vegetable Species	LED Spectral Quality/ Intensity/Photoperiod	Parameters Studied
Chinese Broccoli (*Brassica alboglabra* Bailey)	LED spectral quality (indoors and greenhouse)	Leaf growth, shoot and root productivity, photosynthetic gas exchanges, stomatal conductance, photosynthetic pigments, photosynthetic light use efficiency
Nai bai (*Brassica chinensis* L.) and Mizuna (*B. juncea* var. japonica)	LED spectral quality (greenhouse)	Leaf growth, shoot and root productivity, photosynthetic gas exchanges, stomatal conductance, photosynthetic pigments

Vegetable Species	LED Spectral Quality/ Intensity/Photoperiod	Parameters Studied
Lettuce (Lactuca sativa L.) – Heat-resistant and heat-sensitive recombinant inbred lines (RILs) – Green- and red-leaf lettuce – Cos lettuce	LED spectral quality and intensity and photoperiod; supplemental LED lighting to natural sunlight (indoors and greenhouse)	Leaf growth, shoot and root productivity, photosynthetic gas exchanges, stomatal conductance, photosynthetic pigments photosynthetic light use efficiency, light interception area, light absorption, photosynthetic capacity, photosynthetic characteristics
Common ice plants (Mesembryanthemum crystallinum)	LED spectral quality (indoors)	Leaf growth, leaf water status, shoot and root productivity, photosynthetic pigment photosynthetic gas exchanges, stomatal conductance, light use efficiency, nitrogen metabolism, nutritional quality
Sweet potato	LED intensity, supplemental LED lighting to natural sunlight (greenhouse)	Leaf growth, photosynthetic pigments photosynthetic gas exchanges, stomatal conductance, light use efficiency
Purslane (Portulaca oleracea L.)	LED spectral quality, intensity, photoperiod, DLI (indoors)	Root morphology, leaf growth, leaf water status, shoot and root productivity, photosynthetic pigments, light use efficiency, nitrogen metabolism, nutritional quality

Light intensity, duration, and cycle (photoperiod) are carefully controlled in vertical farms **to optimise plant growth and energy efficiency**.

These factors are important because plants have evolved specific light requirements that vary between species and throughout their growth stages.

Light intensity is measured in micromoles per square meter per second (μmol/m²/s) and refers to the **amount of photosynthetically active radiation (PAR) reaching the plants.** Different crops require different light intensities:

- Leafy greens like lettuce typically need 200-400 μmol/m²/s
- Fruiting plants like tomatoes may require 400-600 μmol/m²/s or higher

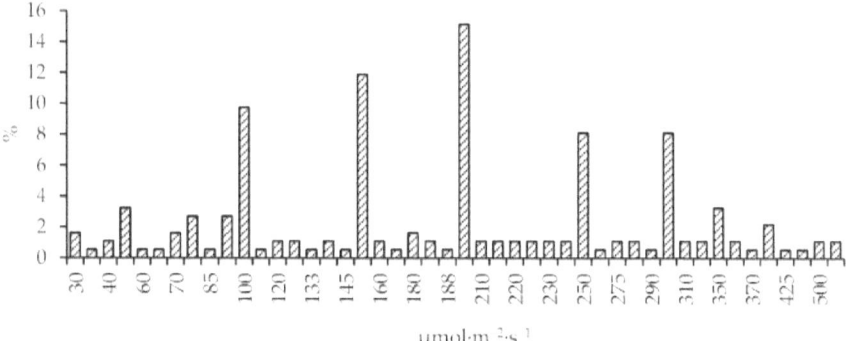

Intensities used in vertical farming systems. n = 185. Source: Najaro et al., 2023

Differential light conditions affect nutrient uptake in multiple crops.
Source: Xu et al., 2021

Light condition		Crop species	Nutrients
Light quality	Red light	Chinese chive (*Allium tuberosum*)	N↑
	Red light	Spinach (*Spinacia oleracea* L.)	Zn↑
	Red light	Gynostemma (*Gynostemma pentaphyllum*)	Ca↑, Fe↑, Zn↑, Cu↑, Se↑
	Red light	Cucumber (*Cucumis sativus* L.)	P↑, K↑, Mn↑, Zn↑
	Blue light	Chinese chive (*Allium tuberosum*)	P↑, K↑
	Blue light	Garlic (*Allium sativum* L.)	N↑, P↑, K↑
	Blue light	Spinach (*Spinacia oleracea* L.)	N↑, Ca↑, K↓, Mg↓
	Blue light	Gynostemma (*Gynostemma pentaphyllum*)	Ca↑, Fe↑, Zn↑, Cu↑, Se↑

Light condition		Crop species	Nutrients
	Blue light	Lettuce (*Lactuca sativa* L.)	Fe↑, Cu↑, Zn↑, Mn↑
	R/B = 4: 1	Celery (*Apium graveolens* L.)	Zn↑
	R/B = 7: 1	Celery (*Apium graveolens* L.)	Se↑
	R/B = 1: 4	Broccoli (*Brassica oleacea* var. *italica*)	Ca↑, Mg↑, P↑, S↑, B↑, Cu↑, Fe↑, Mn↑, Mo↑, Zn↑
	R/B = 1: 1	Mulberry (*Morus alba* L.)	Mn↑, Cu↑
	R/B = 7: 3	Mulberry (*Morus alba* L.)	Zn↑
Light intensity	High light intensity	Red seaweed (*Gracilaria manilaensis*)	NO_3^-↑, NH_4^+↑, PO_4-P↑
	High light intensity	Pakchoi (*Brassica campestris*)	N↑
	High light intensity	Agarophyte (*Gracilaria asiatica*)	NO_3^-↑
	High light intensity	Agarophyte (*Gracilaria asiatica*)	PO_4-P↑
	High light intensity	Thalli (*Hizikia fusiforme*)	P↑
	High light intensity	Cucumber (*Cucumis sativus* L.)	K↑, Ca↑, N↓, Mg↓
	High light intensity	Tomato (*Solanum lycopersicum* L.)	N↓
	Low light intensity	Agavision (*Trifolium pretense* L.)	N↑, P↑
	Low light intensity	Lettuce (*Lactuca sativa* L.)	N↑
Photoperiod	Long photoperiod	Cucumber (*Cucumis sativus* L.)	N↑, P↑, K↑
	Long photoperiod	Lettuce (*Lactuca sativa* L.)	K↑
	Long photoperiod	Lettuce (*Lactuca sativa* L.)	N↑, P↑, K↑

Light condition	Crop species	Nutrients
Long photoperiod	Poinsettia (*Euphorbia pulcherrima*)	P↑
Long photoperiod	Marigold (*Tagetes patula*)	P↑
Long photoperiod	Red raspberry (*Rubus idaeus* L.)	N

Advanced LED systems can **dynamically adjust intensity** based on the plant's growth stage and species-specific needs, ensuring optimal light levels without wasting energy

Scientific Report

Light Regulation of Horticultural Crop Nutrient Uptake and Utilization
Country: China
Publication Date: September 2021
Main Focus: This study explores how different light conditions, including light quality, intensity, and photoperiod, affect the uptake and utilisation of nutrients in horticultural crops.

Key Findings:

- Different wavelengths of light (red, blue, etc.) influence the uptake of various nutrients differently. For instance, red light enhances the uptake of nitrogen, while blue light improves the absorption of phosphorus and potassium.
- Higher light intensity generally increases the absorption of certain nutrients but can decrease others. Optimal light conditions vary between crops.
- Longer light periods can enhance the absorption of nutrients like nitrogen, phosphorus, and potassium, improving overall plant growth.
- Light regulates nutrient uptake through complex signalling pathways involving photoreceptors and transcription factors like HY5, which affect root development and nutrient transport.

- Proper manipulation of light conditions can improve nutrient use efficiency, reduce the need for chemical fertilizers, and enhance crop yield and quality.

Reference: Xu, J., Guo, Z., Jiang, X., Ahammed, G. J., & Zhou, Y. (2021). Light regulation of horticultural crop nutrient uptake and utilization. *Horticultural Plant Journal* **2021**, 7(5), 367-379.DOI: 10.1016/j.hpj.2021.01.005

The **duration of light exposure**, or **photoperiod**, plays a critical role in plant development [Nájera et al., 2023]:

- **Short-day plants** (e.g., chrysanthemums, strawberries) flower when night length exceeds a critical period
- **Long-day plants** (e.g., lettuce, spinach) flower when night length falls below a critical period
- **Day-neutral plants** (e.g., tomatoes, peppers) flower regardless of night length

Vertical farms use **programmable lighting systems** to manipulate photoperiods, inducing flowering or preventing premature bolting as needed.

Scientific Report

Moving Away from 12:12; the Effect of Different Photoperiods on Biomass Yield and Cannabinoids in Medicinal Cannabis
Country: Australia
Publication Date: 27 February 2023
Main focus: This study examines the impact of varying photoperiods on the biomass yield and cannabinoid concentrations in medicinal cannabis, particularly focusing on three varieties: Cannatonic (high-CBD), Northern Lights, and Hindu Kush (both high-THC).

Key findings:
1. The study found that extending the photoperiod to 14 hours of light (14L) significantly increased the biomass yield in all varieties compared to the standard 12 hours of light (12L).

2. For the Cannatonic variety, the CBD concentration increased by 50-100% under a 14L photoperiod. In contrast, THC concentrations in Northern Lights and Hindu Kush decreased by around 40% under the same conditions.
3. The 14L photoperiod resulted in higher flower biomass yields across all tested lines, but the optimal photoperiod varied depending on the specific cannabinoid focus (CBD vs. THC).
4. The study suggests that a tailored photoperiod schedule can optimize cannabinoid yields, highlighting the need for specific lighting strategies for different cannabis varieties.
5. By adjusting photoperiods, growers can potentially increase the yield and potency of cannabis crops, offering economic benefits and better meeting medicinal demands.

Reference: Peterswald TJ, Mieog JC, Azman Halimi R, Magner NJ, Trebilco A, Kretzschmar T, Purdy SJ. Moving Away from 12:12; the Effect of Different Photoperiods on Biomass Yield and Cannabinoids in Medicinal Cannabis. *Plants* **2023**;12(5):1061. DOI: 10.3390/plants12051061

Photographs of the top of the main flowering stem taken at DAC 67 from the three static treatments (10L, 12L, and 14L). Source: Peterswald *et al.*, 2023

Modern LED systems can provide specific light spectra tailored to plant needs.

Blue light (400-500 nm) promotes compact, leafy growth and is crucial for photosynthesis. **Red light** (600-700 nm) is efficient for photosynthesis and influences flowering. **Far-red light** (700-750 nm) can affect plant morphology and flowering. **Green light** (500-600 nm), while less efficient for photosynthesis, penetrates leaf canopies better

Advanced systems may **adjust the spectrum throughout the day or growth cycle** and provide higher blue light ratios during vegetative growth stages and increased red and far-red light during flowering and fruiting stages.

Daily Light Integral (DLI), measured in mol/m²/day, represents the **total amount of photosynthetically active radiation** (400-700 nm) that plants use for photosynthesis, **received in a 24-hour period**.

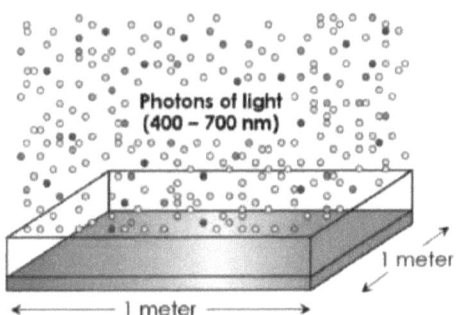

Daily light integral. Source: Michigan State University / *Floriculture & Greenhouse Crop Production* / Erik Runkle, 2006

Calculating DLI Example		
Light intensity values recorded once per hour from midnight to midnight (μmol·m⁻²·d⁻¹)	Average light intensity (μmol·m⁻²·d⁻¹)	Calculated DLI (mol·m⁻²·d⁻¹)
0, 0, 0, 0, 0, 44, 102, 198, 255, 410, 454, 600, 532, 627, 466, 376, 303, 187, 91, 45, 47, 44, 43, 0	201	17.37

An example of hourly light intensity values, the average of those values and the calculated daily light integral (DLI). Source: Michigan State University / *Floriculture & Greenhouse Crop Production* / Erik Runkle, 2006

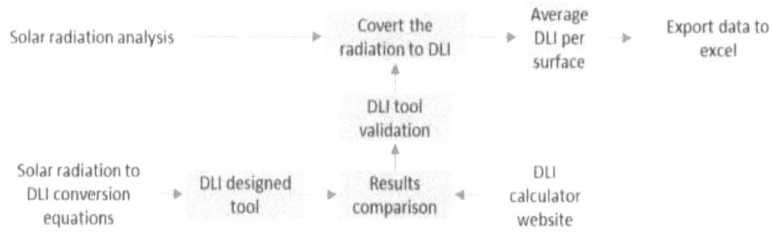

DLI calculation algorithm. Source: Ghazal et al., 2023

Vertical farms can precisely control DLI to match crop requirements:

- **Leafy greens** might need 10-17 mol/m²/day
- **Fruiting crops** often require 20-30 mol/m²/day or more

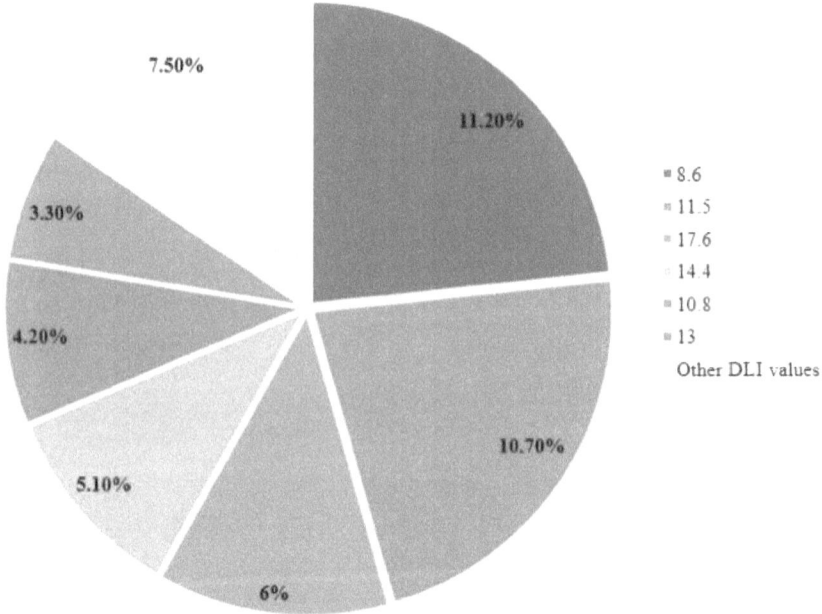

Daily Light Integral (DLI) (mol·m−2·day−1) values used in the studies. n = 215.
Source: Nájera et al., 2023

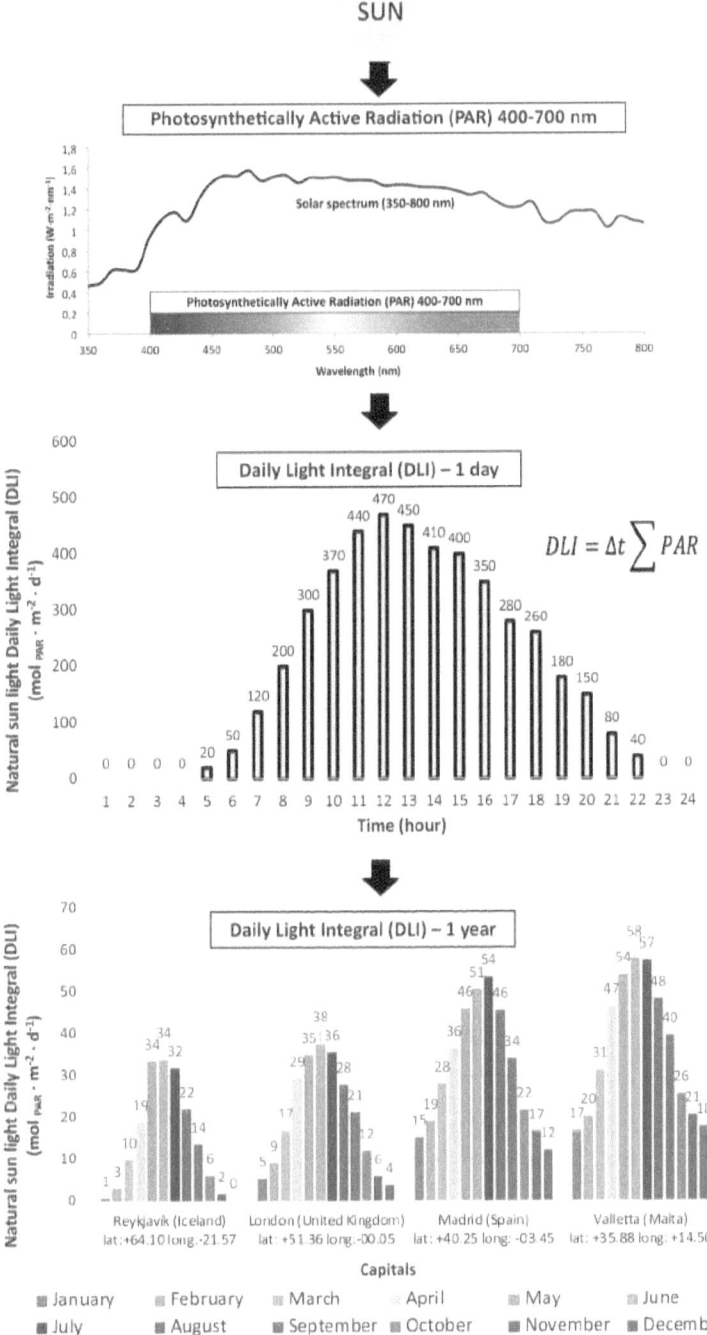

Seasonally averaged DLI values of different cities in the northern hemisphere.
Source: Jung et al., 2024

Scientific Report

Optimal DLI for Indoor Lettuce Cultivation
Country: India
Date: 2023
Main Focus: The study aimed to determine the optimal daily light integral (DLI) for growing iceberg lettuce in a vertical hydroponic system under artificial lighting conditions.

Key Findings:

- The optimal DLI for iceberg lettuce was found to be 11.5 mol m−2 day−1.
- A DLI of 11.5 mol m−2 day−1 significantly increased the fresh weight, dry weight, and leaf area of lettuce.
- Higher DLIs (14.4 mol m−2 day−1) had a negative impact on fresh weight, dry weight, and leaf area.
- The study demonstrated a 60% increase in fresh weight when transitioning from VHS without artificial lights to VHS with artificial lights.

Reference: Kalantari, F., Kalantari, S., Deshkar, A., Hattimare, N., & Jadhav, S. Determination of Optimal Daily Light Integral (DLI) for Indoor Cultivation of Iceberg Lettuce in an Indigenous Vertical Hydroponic System. Scientific Reports **2023**, 13, 36997. DOI: 10.1038/s41598-023-36997-2

Some vertical farms are even experimenting with **"light recipes"** - precise combinations of light intensity, duration, and spectral composition optimised for specific crop varieties and desired characteristics.

Comparison of resource use efficiency of greenhouses and indoor vertical farms. Source: Vatistas et al., 2022

kgFW = Kilogram of fresh weight. kgDW = Kilogram of dry weight. N = Number of air changes. (a) Non-RES use and located outside urban areas, (b) located in peripheral urban areas and use of RES, (c) non-RES use, (d) use of RES.

Resources	GH	VF
Energy	4.5–10.5 kWh kgFW^{-1}	15.6–20.4 kWh kgFW^{-1}
Water	≈10–20 L kgFW^{-1}	1 L kgFW^{-1}
Light	Sunlight and supplementary lighting	AL
Yield	41 kg m^{-2} y^{-1}	150 kg m^{-2} y^{-1}
Land use	365 days per year	365 days per year
Harvests	6–7 per year	8–12 per year
CO2 use	≈14–26 kgCO2 kgDW^{-1}	≈2.1 kgCO2 kgDW^{-1}
CO2 utilisation efficiency	Loses 0.31–0.35 kgCO2 kgFW^{-1}	0.87 (N = 0.01 h^{-1})
CO2 emissions	(a) 0.574 kgCO2 kgFW^{-1} (conventional GH) (b) 0.352 kgCO2 kgFW^{-1} (advanced GH)	(c) 5.7 kgCO2 kgFW^{-1} (conventional VF) (d) 0.158 kgCO2 kgFW^{-1} (green VF)
Pesticide	Use of insect screens for reducing pesticide applications	No use (due to sterilised cultivation environment)

At first glance, the **energy demands** of a vertical farm might seem overwhelming.

Energy demand for operation systems of a greenhouse.
Source: Vatistas et al., 2022

Category	Type	Energy Consumption or Operating Power	GH Characteristics	Location
Heating	Gas	≈383 kWh m^{-2} y^{-1}	Four-span gable roof, double-layer PE film, 1125 m^2, tomatoes,	Simulation, (Saskatoon, Canada)
Heating	Natural gas, coal, heavy oil	≈549 kWh m^{-2} y^{-1}	Venlo-type, double-layer PE film, 81,000 m^2, peppers, h = 3.2 Wm^{-2} °C^{-1}	Leamington (Ontario, Canada)
Heating	Coal	≈100–291 kWh m^{-2} y^{-1}	Gothic roof, plastic-covered, 10,003 m^2	Simulation (5 regions of southern coast of Turkey)
Heating	Gas	≈412 kWh m^{-2} y^{-1}	Venlo-type, glass, 10,000 m^2, h = 5.7 Wm^{-2} °C^{-1}	Simulation (Sweden)
Heating	Gas	≈144 kWh m^{-2} y^{-1}	Venlo-type, glass, 10,000 m^2, h = 5.7 Wm^{-2} °C^{-1}	Simulation (Netherlands)
Cooling	Fan-pads, circulation fans	11.9 kWh total consumption	Glass, multi-span, 2304 m^2	Shanghai (Southeast China)
Cooling	Natural ventilation, fogging system	≈185 kWh m^{-2} y^{-1} (sensible cooling)	Venlo-type, glass, 10,000 m^2, h = 5.7 Wm^{-2} °C^{-1}	Simulation (United Arab Emirates)
Cooling	Natural ventilation, fogging system, heat exchanger, air-cooled chiller	≈700 kWh m^{-2} y^{-1} (dehumidification) ≈844 kWh m^{-2} y^{-1} (sensible cooling)	Venlo-type, glass, 10,000 m^2, h = 5.7 Wm^{-2} °C^{-1}	Simulation (Netherlands)
Lighting	HPS lamps	≈206 kWh m^{-2} y^{-1}	Venlo-type, glass, 10,000 m^2, h = 5.7 Wm^{-2} °C^{-1}	Simulation (Sweden)
Lighting	600 W HPS lamps	90 Wm^{-2} for 48 μmolm^{-2}s^{-1}, 54 Wm^{-2} for 24 μmolm^{-2}s^{-1}	≈75 m^2 compartment in GH	University of Aarhus (Denmark)
Lighting	HPS lamps, LEDs	19,578 kWh (HPS) and 4697 (LEDs) for five months	Glass, two different light treatments in ≈18 m^2 each, tomatoes	West Lafayette (USA)
Ventilation	Fan motor	≈9.7 kWh m^{-2} (from March to October)	Glass, 500 m^2	South-West Greece
Irrigation	Pump water from deep wells	≈83 kWh m^{-2}	26 GHs study, average 2000 m^2, basil	Esfahan (Iran)

h = heat transfer coefficient (Wm−2 °C−1).

Rows of LED lights, HVAC systems, and various pumps and sensors continuously draw power from the grid. As a result, energy becomes one of the largest expenses in vertical farming, often accounting for **25-30% of operating costs.**

Energy demand for operation systems of vertical farming. Source: Vatistas et al., 2022

Category	Type	Energy Consumption	Production Area	Location of VF
Lighting	600 W HPS lamps	1374 kWh m^{-2} y^{-1}	506 m^2	Simulation
Lighting	LED (250 μmol m^{-2} s^{-1} light intensity)	560 kWh m^{-2} y^{-1}	1296 m^2	Simulation (Netherlands)
Lighting	LEDs	26,490 kWh y^{-1} for 60,000 plants' production	N/A	Basement of an urban residential building in Stockholm
Lighting	LED (500 μmol m^{-2} s^{-1} light intensity)	≈1128 kWh m^{-2} y^{-1}	50,000 m^2	Simulation (Sweden)
Cooling	HVAC (forced circulation), fancoil unit, air-cooled chiller	≈86 kWh m^{-2} y^{-1} (Sensible cooling) ≈506 kWh m^{-2} y^{-1} (LED cooling)	50,000 m^2	Simulation (Sweden)
Cooling	HVAC system	≈48 kWh m^{-2} y^{-1}	1891 m^2	Simulation (Riyadh, Saudi Arabia)
Cooling	Chiller	≈404 kWh m^{-2} y^{-1}	1712 m^2	Simulation (Minneapolis, USA, cold-humid climate)
Heating	Natural gas boiler	≈932 kWh m^{-2} y^{-1}	1712 m^2	Simulation (Minneapolis, USA, cold-humid climate)
Heating	HVAC system	≈29 kWh m^{-2} y^{-1}	1891 m^2	Simulation (Seattle, USA)
Dehumidification	HVAC system	≈222 kWh m^{-2} y^{-1}	50,000 m^2	Simulation (Sweden)
Dehumidification	HVAC system	370 kWh m^{-2} y^{-1}	1296 m^2	Simulation (United Arab Emirates)
Irrigation	Pump	≈18 kWh m^{-2} y^{-1}	506 m^2	Simulation
Irrigation	Pump	2190 kWh y^{-1} for 60,000 plants' production	N/A	Basement of an urban residential building in Stockholm

Vertical farms are exploring additional uses for their lighting systems beyond plant illumination. Some operations are **using the heat generated by LED lights for climate control purposes**.

By strategically positioning lights and employing heat-conductive materials, farms can **redirect this thermal energy** to warm growing spaces, potentially reducing the demand on heating systems.

In cooler climates, this excess heat may be used to **warm adjacent spaces** such as offices, improving overall energy efficiency.

Scientific Report

Thermal Management and Energy Efficiency Analysis of Planar-Array LED Water-Cooling Luminaires in Vertical Farming Systems for Saffron
Country: China 🚩
Publication Date: 30 September 2023
Main focus: This study investigates the design, thermal management, and energy efficiency of planar-array LED water-cooling luminaires specifically developed for vertical farming systems cultivating saffron.

Key findings:
- The study found that LED water-cooling luminaires, particularly those without power supplies and dimming control systems (WCL-PS), provided optimal thermal management, preventing heat from affecting saffron corm development.
- The photosynthetic photon flux efficacy (PPFE) for white-LED luminaires was higher at 2.75 µmol J^{-1} compared to blue-red LED modules, which had PPFE values of 2.74 µmol J^{-1} for blue LEDs and 1.65 µmol J^{-1} for red LEDs.
- Additionally, the white-LED luminaires achieved better temperature field distribution and energy efficiency, making them more suitable for saffron cultivation in vertical farming systems.

Reference: Wong, C. E., Teo, Z. W. N., Shen, L., & Yu, H. Seeing the lights for leafy greens in indoor vertical farming. *Trends in Food Science & Technology* **2020**, 106, 48-63. DOI: 10.1016/j.tifs.2020.09.031

The **integration of renewable energy sources** is becoming increasingly common in vertical farms. **Solar panels** on the roof can offset a significant portion of the farm's energy needs during daylight hours. Some operations are exploring the **use of wind turbines** or even **geothermal systems**, depending on their location. The goal for many is to **achieve net-zero energy use**, where the farm produces as much energy as it consumes over the course of a year.

> Advanced control systems use machine learning algorithms to continuously optimise energy use across all farm systems. These AI managers can **predict energy needs** based on factors like weather forecasts, crop growth stages, and even electricity pricing, adjusting operations to minimise costs and maximise efficiency.

Some farms are taking energy management to the next level by **participating in demand response programs with local utilities**. During peak demand periods, they can temporarily reduce their energy consumption, earning credits or payments while helping to stabilise the broader power grid. This not only improves the farm's bottom line but also contributes to community energy resilience.

The pursuit of energy efficiency in vertical farming presents a **technical challenge** that encourages a reevaluation of our approach to energy and food production. It motivates the **development of systems that are both productive and sustainable** in the long term.

Temperature

Temperature control is about keeping plants comfortable but also it's an intricate **balance of precision that can make or break a crop**.

> Unlike traditional agriculture, where farmers are dependent on weather patterns, vertical farms offer an unprecedented level of control over the thermal environment. **Day and night temperatures** play crucial roles in plant growth, development, and metabolism.

At its core, temperature management in vertical farms is about **understanding and manipulating plant physiology**. Each crop has its own ideal temperature range, a sweet spot where photosynthesis, respiration, and nutrient uptake operate at peak efficiency. But it's not as simple as setting a thermostat and walking away. For example, vertical farmers develop **Crop-Specific Temperature Protocols** for different crops:

- Leafy greens like lettuce might thrive at 18-24°C (64-75°F). For example, the optimal temperature for basil and lettuce growth depends on the growth stage and specific cultivar. Generally, cooler temperatures are ideal, though these plants can endure some degree of heat stress [Farhangi et al., 2023]
- Tomatoes often require warmer temperatures, around 21-29°C (70-85°F)
- Some herbs, like basil, prefer consistently warm temperatures around 25-30°C (77-86°F)

Plants, like humans, have their own **circadian rhythms**, responding to daily temperature fluctuations in a process known as **thermoperiodism**.

Experienced vertical farmers leverage this knowledge, creating **day-night temperature differentials (DIF)** to sculpt plant growth. A positive DIF, with warmer days and cooler nights, might be used to encourage stem elongation in certain crops. Conversely, a negative DIF can produce more compact plants such as leafy greens.

> Advanced vertical farms employ **micro-zoning**, essentially creating multiple climate bubbles within a single facility. This allows for the **simultaneous cultivation of crops with varying temperature needs** or **the nurturing of a single crop through different growth stages**, each with its own thermal requirements.

Temperature control in vertical farms is linked to humidity management through the concept of **Vapour Pressure Deficit (VPD)**. By precisely controlling both temperature and humidity, farmers can maintain the ideal VPD for each crop, optimising transpiration and nutrient uptake. It's a delicate balance—too high a VPD can stress plants, while too low can invite fungal diseases.

Innovation in temperature control extends to the roots as well. Some systems employ **separate temperature control for the root zone**, recognizing that optimal root temperatures often differ from ideal leaf temperatures. For example, techniques like nutrient film technique (NFT), allow for precise root temperature control. This level of precision can enhance nutrient absorption and overall plant health.

Scientific Report

Enhancing Productivity and Improving Nutritional Quality of Subtropical and Temperate Leafy Vegetables in Tropical Greenhouses and Indoor Farming Systems
Country: Singapore
Publication Date: 21 March 2024
Main focus: The study investigates methods to enhance the productivity and nutritional quality of subtropical and temperate leafy vegetables in tropical greenhouses and indoor farming systems, focusing on techniques such as root-zone temperature manipulation, deficit irrigation, and LED lighting.

Key findings:
- The research demonstrates that cooling the root zone temperature can significantly enhance the yield and nutritional quality of vegetables, with arugula showing a 15% increase in total phenolic compounds under optimal conditions.
- Additionally, deficit irrigation techniques improved water use efficiency without compromising yield, while optimal LED lighting conditions increased the nutritional value and biomass of crops like lettuce and common ice plants.

Reference: He, J. Enhancing Productivity and Improving Nutritional Quality of Subtropical and Temperate Leafy Vegetables in Tropical Greenhouses and Indoor Farming Systems. *Horticulturae* **2024**, *10*, 306. DOI: 10.3390/horticulturae10030306

Since energy efficiency is a constant concern in vertical farming, and temperature control is no exception. Some farms use computational fluid dynamics to **model and optimize airflow**, ensuring every corner of the farm maintains the convenient temperature without wasting energy.

> Temperature manipulation isn't always about creating ideal conditions—sometimes, a little stress can be beneficial. **Brief periods of heat stress** can boost antioxidant content in some crops, while **controlled cold stress** can enhance sweetness by triggering sugar accumulation.

Vertical farms can precisely control **the hardening process for outdoor crops**. By gradually exposing plants to cooler temperatures, farmers can better prepare them for outdoor conditions, a precision that is difficult to achieve in traditional agriculture.

Humidity

Humidity plays a subtle but essential role in influencing all aspects of plant growth and health. Although it may not be as prominent as factors like lighting and nutrition, **effective humidity management** is fundamental for any vertical farm. The goal is to create an environment where every drop of water vapour supports optimal growth.

> Most crops thrive in what we might call a **"Goldilocks zone" of humidity** — not too high, not too low, but just right. Typically, this sweet spot falls **between 50% and 70% relative humidity (RH)**.

However, like many aspects of vertical farming, **one size does not fit all.** Leafy greens, for instance, often prefer the higher end of this range, luxuriating in humidity levels of 60-70% RH. Fruiting plants like tomatoes, on the other hand, tend to favour slightly drier conditions, around 50-60% RH.

Going outside this optimal range can cause multiple problems. High humidity promotes fungal growth, leading to issues like powdery mildew and botrytis. It can also slow transpiration, the plant's natural cooling mechanism, potentially causing nutrient deficiencies and slower growth. Conversely, low humidity stresses plants, causing them to absorb water faster than they can use it efficiently. This increases water consumption and can result in leaf burn and wilting.

While relative humidity has long been the go-to metric for farmers, vertical farming is exploiting one more metric – **Vapour Pressure Deficit (VPD)**. VPD measures the difference between the amount of moisture in the air and how much moisture the air can hold when saturated. It's a more nuanced way to understand how humidity affects plant transpiration.

Low VPD, linked to high humidity, can significantly **reduce transpiration**, thereby impeding nutrient uptake. Conversely, high VPD, indicating low humidity, can lead to excessive water loss and stress for plants. The key is to find the right balance, which varies throughout a plant's life cycle. For instance, seedlings and young plants generally thrive with lower VPD (higher humidity), while mature plants can usually tolerate higher VPD levels.

Dehumidification systems, ranging from refrigerant-based units to desiccant wheels, work to **remove excess moisture from the air**. On the flip side, **humidification systems** like ultrasonic misters and high-pressure foggers can add moisture when needed.

But it's not just about adding or removing water vapour. **Air circulation** plays a crucial role, with horizontal airflow fans preventing stagnant microclimates and vertical air movement ensuring uniform humidity distribution. Some advanced farms even employ climate zoning, creating separate humidity environments for different crops or growth stages within the same facility.

Vertical farmers must **adapt their humidity strategies to various crops.** Leafy greens thrive in higher humidity, which helps maintain their crisp texture. Tomatoes require more nuanced management, with careful humidity control during flowering to ensure proper pollination. Strawberries prefer slightly drier conditions during fruiting to ward off mould, while microgreens demand high humidity during germination, gradually tapering off as they mature.

It's worth noting that **humidity control can be energy-intensive**, a fact not lost on an industry striving for sustainability. Forward-thinking farms are addressing this challenge head-on, implementing **heat recovery systems** that capture and reuse energy from dehumidification processes.

Smart scheduling of lighting and irrigation helps manage humidity passively, while **integration with Heating, Ventilation and Air Conditioning (HVAC) systems** allows for holistic climate control, optimising energy use across all environmental parameters.

As vertical farming technology improves, there are advances in humidity management. **Nanotechnology sensors** offer ultra-precise humidity monitoring, while **biomimetic systems** copy natural forest humidity regulation.

Air Circulation

In the precisely controlled environment of a vertical farm, air isn't just something plants breathe—it's a dynamic force that shapes every aspect of plant growth.

Air circulation, often overlooked in discussions of lighting and nutrition, is playing a crucial role in creating the proper ecosystem for plant development.

In nature, disappearance of wind would be catastrophic for plant life. In a vertical farm, the absence of natural air movement could spell disaster. This is where **engineered air circulation** steps in, mimicking and improving upon nature's own systems.

At its most basic level, air circulation in vertical farms serves to **distribute heat evenly throughout the growing space**. Without proper air movement, hot and cold spots can develop, creating microclimates that may stress plants or lead to inconsistent growth. But the benefits of good air circulation extend far beyond temperature regulation.

Consider the hidden threat of **trapped air pockets**. These areas of still air can become breeding grounds for pests and pathogens, threatening entire crops if left unchecked. By ensuring **constant air movement**, vertical farmers create an environment that's inhospitable to many common agricultural pests. Fungal spores, which thrive in still, humid conditions, find it difficult to settle and germinate when there's a constant gentle breeze.

But air circulation does more than just ward off potential threats—it actively **contributes to plant health and vigour**. As air moves across leaf surfaces, it helps to strengthen plant stems and tissues. This process, known as **thigmomorphogenesis**, mimics the natural stress of wind, encouraging plants to develop stronger cellular structures. The result? Plants that are stronger and can better support their fruits or vegetables.

Perhaps one of the most critical functions of air circulation in vertical farming is its role in **carbon dioxide distribution.** Plants, as we know, require CO_2 for photosynthesis. In an enclosed environment like a vertical farm, CO_2 levels can quickly drop around the leaves as plants engage in photosynthesis. Effective air circulation ensures a steady supply of CO_2-rich air to all plants, maximising photosynthetic efficiency and, by extension, growth rates and yield.

As mentioned, **humidity management**—another cornerstone of successful vertical farming—is closely tied to air circulation. As plants transpire, they release water vapour into the air immediately surrounding their leaves. Without proper air movement, this humid microclimate can inhibit further transpiration, potentially leading to nutrient deficiencies and increased susceptibility to fungal diseases. Good air circulation removes extra humidity, keeping the conditions ideal for plant transpiration and nutrient uptake.

But how do vertical farmers achieve this balance of air movement? The answer lies in a carefully designed **system of fans and ducts**. Horizontal airflow fans, strategically placed throughout the growing area, create a gentle but constant breeze across plant canopies. Vertical air movement is often managed through carefully designed ductwork, ensuring that air circulates not just within each growing level, but between levels as well.

Advanced vertical farms are taking air circulation to new heights with **computational fluid dynamics (CFD) modelling**. This advanced tool allows farmers to visualise and optimise airflow patterns, ensuring that every plant receives the proper amount of air movement. Some systems even incorporate **variable-speed fans** that adjust their output based on plant growth stage or environmental conditions, providing dynamic air circulation.

> Modern vertical farms often use **air movement as a tool for fine-tuning temperature and humidity levels**. For example, increasing air circulation can help to reduce humidity without the need for energy-intensive dehumidification systems. Similarly, strategic air movement can help distribute cooling or heating more efficiently throughout the growing space.

Emerging technologies like **biomimicry-inspired air distribution systems** promise to recreate the complex air movement patterns found in natural ecosystems.

Carbon Dioxide Augmentation

In vertical farms, carbon dioxide levels are often augmented to **enhance photosynthesis and plant growth**.

Often referred to as the "aerial fertiliser," CO2 augmentation in vertical farming **enhances plant productivity**, highlighting the potential for high-yield, sustainable agriculture. CO2 augmentation leverages the fundamental process of photosynthesis.

In nature, plants use atmospheric CO2 levels of around 400 parts per million (ppm). However, in the controlled environment of a vertical farm, **CO2 levels can be increased to significantly boost plant growth**.

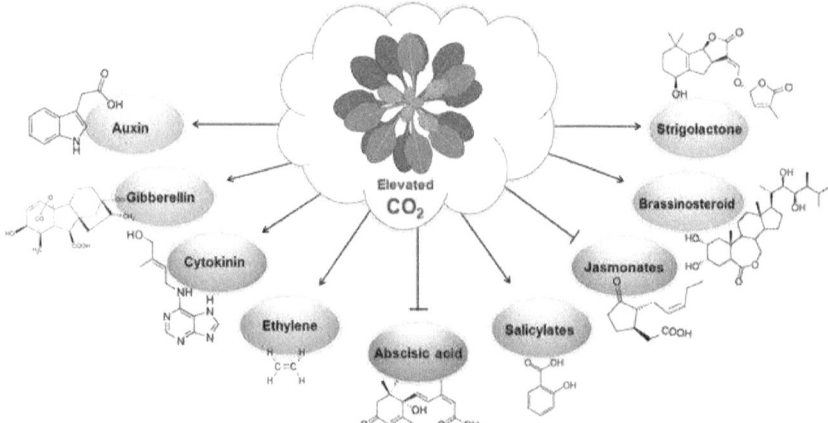

Higher levels of CO2 usually increase the production and activity of most plant hormones like auxins, gibberellins, cytokinins, ethylene, salicylic acid, brassinosteroids, and strigolactones. However, the pathways for abscisic acid and jasmonate are often reduced. These changes in hormone levels due to elevated CO2 can affect plant growth and how plants respond to stress, influencing many physiological and developmental processes.
Source: Ahammed et al., 2021

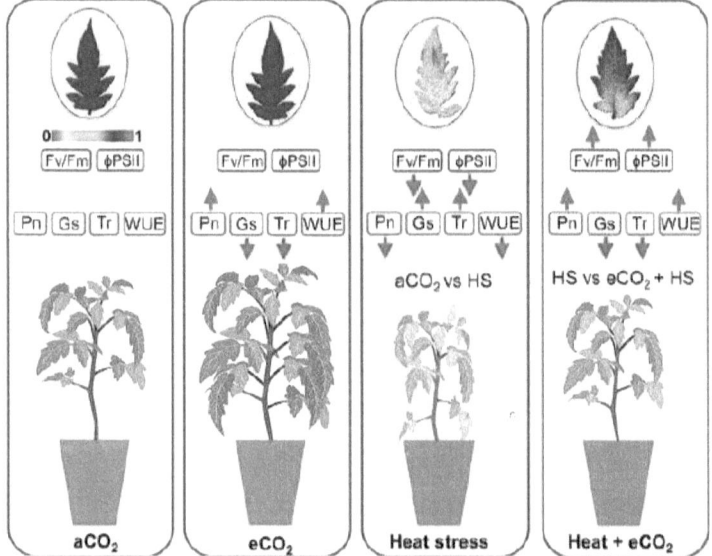

Interactive effects of elevated CO2 and heat stress on leaf gas exchange, chlorophyll fluorescence and biomass accumulation. In C3 plants, elevated CO2 triggers net photosynthetic rate (Pn), but reduces stomatal conductance (Gs) and transpiration rate (Tr), leading to increased water use efficiency (WUE) and biomass accumulation. Heat stress (HS) drastically declines photosynthetic performance, by inhibiting the Pn, Fv/Fm, φPSII and WUE; however, elevated CO2 mitigates the deleterious effects of heat stress on plant photosynthesis and biomass accumulation. Source: Ahammed et al., 2021

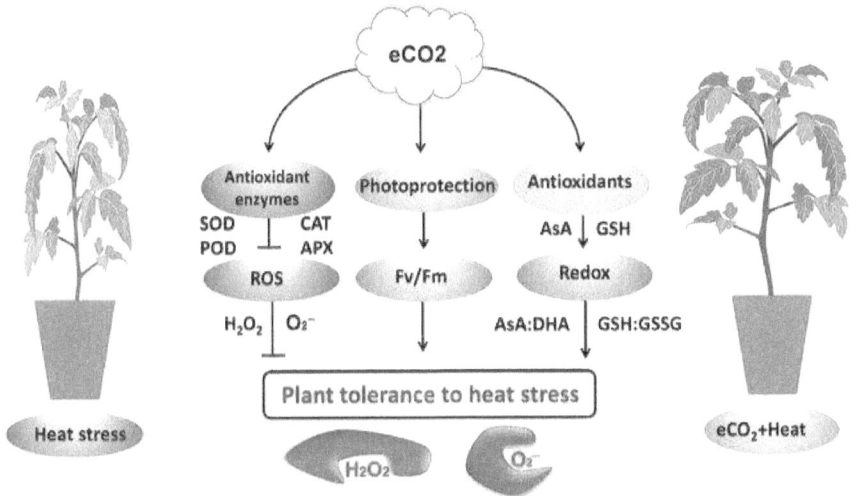

ABA-independent thermotolerance

Elevated CO2 induces thermotolerance by activating the antioxidant defense and redox homeostasis in an abscisic acid (ABA)-independent process. eCO2 elevated CO2, SOD superoxide dismutase, CAT catalase, POD peroxidase, APX ascorbate peroxidase, ROS reactive oxygen species, H2O2 hydrogen peroxide, O2·- superoxide, Fv/Fm the maximal photochemical efficiency of photosystem II, AsA ascorbic acid, DHA dehydroascorbate, GSH reduced glutathione, GSSG oxidized glutathione. Source: Ahammed et al., 2021

Vertical farmers typically aim for CO2 levels between 1,000 and 1,500 ppm, which is three to four times higher than normal levels. At these elevated concentrations, **plants experience a dramatic boost in photosynthetic efficiency**. The results are nothing short of remarkable: faster growth rates, increased biomass production, and often, enhanced crop quality. It's as if we're giving plants a shot of espresso, accelerating their natural processes to levels unattainable in traditional agriculture.

> There's a delicate balance to strike - a Goldilocks zone where CO2 levels are just right. **Raising concentrations too much can cause less benefit or even harm**. Some crops might lose nutrients, and in extreme cases, high CO2 levels can be toxic to plants.

Each crop has its own CO2 sweet spot. Leafy greens like lettuce might thrive at 1,000-1,200 ppm, while tomatoes and peppers could benefit from levels up to 1,500 ppm. Some flower crops even respond well to concentrations as high as 2,000 ppm. Finding these optimal ranges requires a deep understanding of plant biology and a willingness to experiment and fine-tune.

Like any powerful tool, CO_2 augmentation comes with its challenges. The systems can be **expensive** to implement and maintain, especially for larger operations. **Safety** is a primary concern, as high CO_2 levels can be dangerous for humans working in the farm. There's also **the environmental aspect** to consider - care must be taken to prevent CO_2 leakage into the broader environment, and in some regions, enrichment practices may be subject to regulations.

Implementing CO_2 augmentation in a vertical farm is no small feat, it means that adding CO_2 to a vertical farm is a **challenging and significant achievement**. It requires advanced equipment and careful planning. CO_2 can be introduced from compressed tanks, generated on-site by burning propane or natural gas, or even captured from industrial processes as a sustainable option. Distribution systems, ranging **from perforated tubing to overhead dispensers**, ensure that this gas reaches every plant evenly.

One of the most promising routes are the experiments with **photobioreactors** - systems that cultivate algae alongside traditional crops. These algae consume CO_2 and produce oxygen, creating a symbiotic relationship that could revolutionise the way we think about atmospheric management in controlled environments.

AI and Internet of Things (IoT)

Internet of Things (IoT) in vertical farming **enables real-time monitoring and control** of environmental conditions, resource usage, and plant health, thereby increasing efficiency and reducing operational costs.

On more advanced level, artificial intelligence (AI) in vertical farming **optimises farming practices** through data analysis, predictive modelling, and automation, leading to enhanced crop yields and improved decision-making processes

IoT applications in vertical farming. Source: Siregar *et al.*, 2022

Area	Application	Approach Method
Hydroponics, Tomato plant growth	Developed using Arduino, Raspberry Pi3, and TensorFlow.	Deep Neural Networks
Urban farming	NB-IoT sensor network (deployed in balconies of two multistory building structures)	Fuzzy logic
Urban farming decision support framework	DSS to online database captured by IoT technologies and robotic machines, it is promising to achieve a high level of automation in the field of urban agriculture	Decision support systems (DSS) based on modelling and simulation have been developed to assess farming systems
Plant Factory Artificial Light (PFAL) management system	Use of artificial intelligence (AI) with a database, Internet of Things (IoT), light-emitting diodes (LEDs), and phenotyping unit	AI-based smart PFAL management system
Plan Factory	Greenhouse environmental monitoring, develop complex mathematical models to minimise energy input or use solar or wind energy	Controlled environment agriculture (CEA)
Smart indoor farming—secure and self-adapting	The important element for such solutions is a cloud, IoT, and robotics-based smart farming framework.	AgroRobot and Indoor Farming Support as a Service (IFSaaS)

Area	Application	Approach Method
Real-Time Greenhouse Environmental Conditions	Determination of nutrients needed for plant growth, such as nitrogen (N), phosphorus (P), and potassium (K) in soil or water, is the key to vertical or closed crop cultivation with IoT	Clustering quantity using the K-means method and a prediction approach using the Self-Organizing Map (SOM) method to enhance the device capacity and real-time analytics
CPS/IoT Ecosystem: Indoor Vertical Farming System	A prototype is a service-oriented platform distributed over three scopes of operation: cloud, fog, sensor/actuator	Smart agricultural systems using CPS/IoT infrastructure and offering Infrastructure-as-a-Service (IaaS) and Experiment-as-a-Service (EaaS) for smart farming
Digital Twins for Vertical Farming	Design science research paradigm, aiming at the joint creation of physical and digital layers of IoT-enabled structures for vertical farming	A digital twin reference model for IoT-enabled structures of vertical farming
Automatic vertical hydroponic farming	The design and implementation of automated vertical hydro farming techniques with IoT platforms, and their analytics will be conducted using big data analytics	Automatic robotic system design and development
IoT-Enabled in Smart Vertical Farming	Application of IoT-Enabled Smart Agriculture in Vertical Farming	The web-based application can be used to analyse and monitor the light, temperature, humidity, and soil moisture of the vertical farming stacks
Review adoption of vertical gardens (VG) and/or vertical farms (VF)	Automating sustainable vertical gardening systems by using the IoT concept in smart cities toward smart living	Literature review
Indoor Vertical Farming	Build a system to monitor the soil moisture and control water content	Automatic a system, which consists of the Internet of Things

Studies in the use of AI for the development of intelligent vertical agriculture.
Source: Siregar et al., 2022

Model ML/Algorithms	Approach	Application	Crops/Area	Observed Features
Genetic Algorithm and Job-Shop Scheduling	Presenting an efficient method based on the genetic algorithm developed to solve the proposed scheduling problem	Indoor vertical farming	Fruits	Control and increase food production and predict harvest time
RNG k-epsilon model	RNG k is implemented to consider the impact of air pressure and barriers in the computing domain	Indoor vertical farming	Vegetable plant	A three-dimensional numerical model for optimizing airflow and heat transfer in a vertical farm space system taking into account carbon dioxide consumption, and oxygen production.
Fuzzy Logic	Fuzzy logic handles certainty and fuzzy evaluation based on the distance from the mean solution (EDAS) method assists in the system evaluation decision-making process.	Hydroponics system in vertical farming	Planting system without soil	Number of crops type, production volume, attractiveness, sustainability, flexibility, workforce requirement, stock-out cost, transportation cost, and investment cost
Fuzzy Logic, WEDBA (Weighted Euclidean Distance Based Approximation) & MACBETH (Measuring Attractiveness by a Categorical Based Evaluation Technique)	The WEDA and MACBETH methods were used to rank three smart farming alternatives in urban areas	Smart system in hydroponic vertical farming	Planting system without soil	Venture capital attractiveness, effective manufacturing process, workforce requirement, security, space requirement, R&D capabilities, expansion opportunities, investment, and maintenance cost
Integer Linear Program (ILP)-Crop Growth Planning Problem (CGPP)	Present four mathematical models for planning the growth of crops in a vertical farming system, which are strengthened using variable fixing and valid inequalities.	Vertical farm	Leaf vegetables	Machine scheduling and configuration
Mixed-Integer Linear Programming (MIP)	Three approaches using polynomial, pseudo, and hybrid variables (polynomial end pseudo)	Vertical farming elevator energy minimization problem (VFEEMP)	Vertical agriculture energy source	Driving energy in vertical farms

Model ML/Algorithms	Approach	Application	Crops/Area	Observed Features
Computer vision—Machine learning	Viola-Jones algorithm and Haar-like feature extraction method for the machine learning	Detect spot disease in tomatoes	Telemetry vertical farming	Detection of spot disease in tomatoes is designed using 377 images of infected tomatoes
Multi-criteria decision-making (MCDM) and Pythagorean fuzzy set (PFS) Pythagorean Fuzzy Choquet Integral (PFCIμ)	Multi-criteria decision-making (MCDM) framework to assess the VF systems. A novel Pythagorean fuzzy set (PFS) with Choquet Integral model integrated is recommended for VF technology evaluation	Vertical farming feasibility evaluation framework	Comparative study of agricultural vertical land	Evaluate the urban farming framework to choose the right strategy and change the strategy
Fuzzy Logic	The fuzzy logic control will be based on the state of charge (SoC) of each node. A wireless sensor network (WSN) will provide two-way communication between nodes and coordinators.	Development and design of power generation and distribution optimised for vertical farming.	Power generation and distribution for vertical farming	New renewable energy in vertical farming
Support Vector Machine (SVM), Decision Tree (DT), and Neural Network (NN)	Three categories of AI models commonly used in soil management and agricultural production to enable smart farming to be introduced	Multiphonics Vertical Farming(MVF) system	Leaf vegetable	The study discusses how AI is adopted in soil management and MVF for tasks including classification, detection, and forecasting
Feed Forward Neural Network	Regression type feed-forward deep learning a neural network has been utilised	Greenhouse	Tomatoes	The growth of the plants is checked every 24 h and based on the growth, the necessary conditions are provided for the target growth
Machine learning-computer vision	Automatic method for extracting phenotype features, based on CV, 3D modeling and deep learning. From the extracted features, height, weight, and leaf area were predicted and validated with ground truths obtained manually	Vertical farms with artificial lighting (VFAL)	Vegetables	Methods for vision-based plant phenotyping in indoor vertical farm under artificial lighting. This method combines 3D plant modeling and deep segmentation of higher leaves, for 25–30 days, associated with growth

Scientific Report

Vertical Farming Perspectives in Support of Precision Agriculture Using Artificial Intelligence: A Review

Country: Indonesia
Publication Date: 8 September 2022
Main focus: This research reviews the application of artificial intelligence (AI) in vertical farming to support precision agriculture, addressing challenges, technological trends, and future opportunities in this innovative agricultural method.
Key findings:
- The study highlighted several key benefits and applications of AI in vertical farming. For example, the use of IoT in vertical farming can increase productivity by up to 30%.
- Additionally, AI-driven vertical farming systems can reduce water usage by up to 95% compared to traditional farming methods.
- The integration of AI and IoT improves yield predictions and disease monitoring accuracy by 20-25%, showcasing significant advancements in the efficiency and sustainability of vertical farming.

Reference: Siregar, R.R.A.; Seminar, K.B.; Wahjuni, S.; Santosa, E. Vertical Farming Perspectives in Support of Precision Agriculture Using Artificial Intelligence: A Review. *Computers* **2022**, *11*, 135. DOI: 10.3390/computers11090135

Scientific Report

Technological Trends and Engineering Issues on Vertical Farms: A Review

Country: South Korea
Publication Date: 15 November 2023
Main focus: This review explores the current technological landscape and engineering challenges in vertical farming, highlighting advancements in sensing technologies, monitoring systems, and the role of artificial intelligence (AI) in optimising farming processes.

Key findings:

- The global vertical farming market reached USD 5.6 billion in 2022 and is expected to surpass USD 35 billion by 2032, driven by urbanization and the increasing demand for sustainable agriculture.
- Advanced technologies such as sensors, automation, and AI are becoming integral in vertical farming, allowing for precise control of

- environmental factors like light, temperature, and nutrient levels, which in turn optimizes crop yield and quality.
- Vertical farms are more water-efficient than traditional farming methods, utilizing hydroponic and aeroponic systems that minimize water usage and nutrient wastage, making them a sustainable option for food production.
- Vertical farming systems are capable of growing a wide range of crops, including leafy greens, herbs, berries, and even non-edible plants like ornamental flowers, within controlled environments, thereby enhancing food security and supply stability.
- Despite its benefits, vertical farming faces challenges such as high initial setup costs, energy consumption, and the need for technological advancements in sensor accuracy and automation. Addressing these challenges is crucial for the widespread adoption and economic viability of vertical farming systems.

Reference: Kabir, M.S.N.; Reza, M.N.; Chowdhury, M.; Ali, M.; Samsuzzaman; Ali, M.R.; Lee, K.Y.; Chung, S.-O. Technological Trends and Engineering Issues on Vertical Farms: A Review. *Horticulturae* **2023**, *9*, 1229. DOI: 10.3390/horticulturae9111229

Scientific Report

Empowering Vertical Farming through IoT and AI-Driven Technologies: A Comprehensive Review

Country: India
Publication Date: 23 July 2024
Main focus: This review explores the application of IoT and AI technologies in vertical farming, emphasising their roles in improving crop management, disease detection, yield prediction, and overall efficiency.

Key findings: The integration of AI and IoT can significantly enhance crop yield predictions, with some systems achieving over 95% accuracy in predicting lettuce growth. The use of IoT sensors and AI algorithms also led to a 20.4% reduction in power consumption during temperature and water level regulation and an 82.1% reduction during light regulation.

Reference: Rathor, A. S., Choudhury, S., Sharma, A., & Nautiyal, P. Empowering Vertical Farming through IoT and AI-Driven Technologies: A Comprehensive Review. *Heliyon* **2024**, 10, e34998. DOI: 10.1016/j.heliyon.2024.e34998

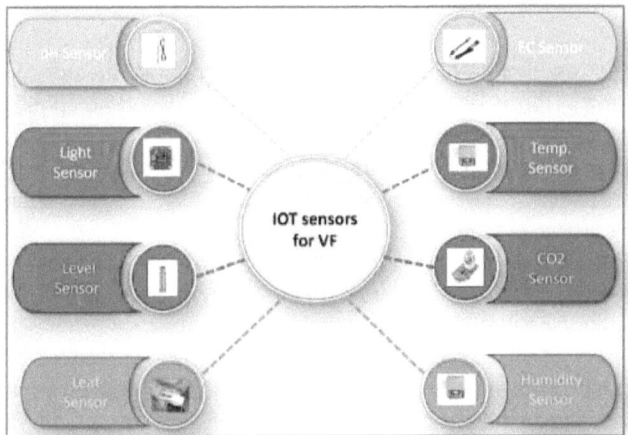

IoT sensors for VF. Source: Rathor *et al.*, 2024

There are three main areas in vertical farming systems, where Internet of Things (IoT) and AI monitoring can help the most - **precision crop management, pest and disease monitoring** and **resource optimization.**

Precision Crop Monitoring

Precise control over environmental conditions such as light, temperature, and nutrients is the core task. Technologies adjust these factors based on real-time data, optimising growth conditions for each stage of the crop's lifecycle. For example, by fine-tuning lighting schedules to enhance photosynthesis and growth.

Scientific Report

Crop Growth Monitoring System in Vertical Farms Based on Region-of-Interest Prediction
Country: South Korea
Publication Date: 30 April 2022
Main focus: The study focuses on developing a crop growth monitoring system for vertical farms using a novel method called pseudo crop mixing to enhance the performance of AI-based monitoring systems.

Key findings:

- The research presents an image-based crop growth monitoring system for vertical farms using a novel method called pseudo crop mixing to improve the performance of crop area prediction models.
- The study found that the proposed model achieved a mean average precision (mAP) of 76.9%, which is 12.5% better than the existing methods.
- The pseudo crop mixing technique helps in increasing the diversity of training data by generating synthetic images from unlabeled data, enhancing model robustness and accuracy.
- The system automates crop growth monitoring, allowing for precise crop condition evaluations and reducing the need for human intervention, which can significantly increase the productivity and reduce the operating costs of vertical farms.

Reference: Hwang, Y.; Lee, S.; Kim, T.; Baik, K.; Choi, Y. Crop Growth Monitoring System in Vertical Farms Based on Region-of-Interest Prediction. *Agriculture* **2022**, *12*, 656. DOI: 10.3390/agriculture12050656

Examples of vertical farms environment: (**a**) a vertical farm cultivation environment; and (**b**,**c**) images of crop growth monitoring. Source: Hwang *et al.*, 2022

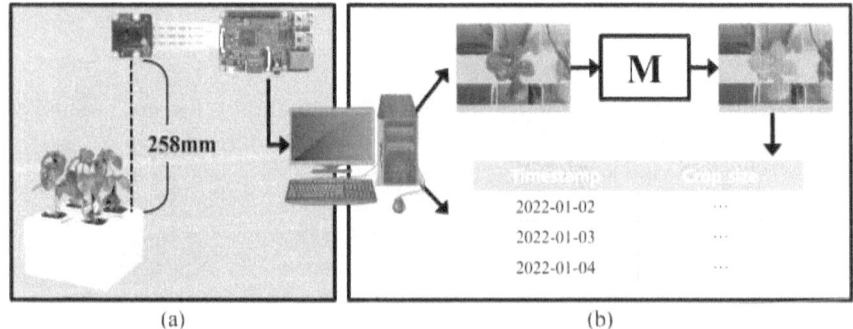

Framework of the crop monitoring system: (**a**) the data collection procedure; and (**b**) crop size estimation. The image captured at the top of the growth tray is transferred to the server through the secure copy protocol (SCP), and the server can track the growth stage of the crop by calculating and recording the occupied area of the crop in the image.
Source: Hwang *et al.*, 2022

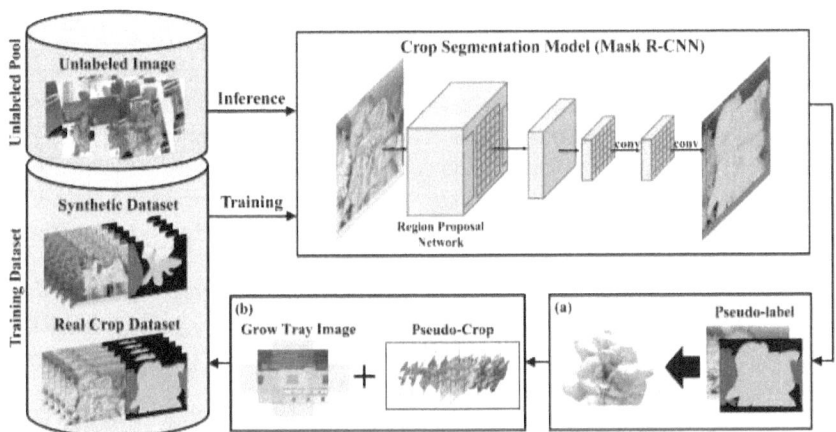

The SC-Mix framework consists of two modules: (a) the pseudo-crop generation (PCG) module, which generates a pseudo-crop from an unlabeled image by converting the pseudo-label of the monitored crop only; (b) the synthetic-image generation (SIG) module, which synthesises vertical farm images using pseudo-crops. These synthetic images are added to the training dataset to re-train the crop segmentation model.
Source: Hwang et al., 2022

Pest and Disease Detection

Sensors and cameras can **detect early signs of pests and diseases in crops.**

These systems analyse visual data to identify abnormalities, allowing farmers to address issues quickly. Early detection helps prevent the spread of diseases and reduces the need for pesticides.

Scientific Report

Monitoring Root Rot in Flat-Leaf Parsley via Machine Vision by Unsupervised Multivariate Analysis of Morphometric and Spectral Parameters

Country: United Kingdom
Publication Date: 19 February 2024
Main focus: This study investigates the use of machine vision for monitoring root rot in flat-leaf parsley grown in a hydroponic vertical farming system, focusing on early detection of disease through morphometric and spectral analysis.

Key findings:
- The study utilized a 3D multispectral scanner to monitor structural and spectral changes in the plant canopy, identifying clear distinctions between healthy and diseased plants at 4-7 days post-inoculation (DPI) using combined morphometric and spectral features.
- Morphometric features like digital biomass and plant height were significant indicators of plant health, showing stunted growth in infected plants.
- Spectral features such as Green Leaf Index (GLI) and Normalised Difference Vegetation Index (NDVI) were higher in healthy plants, whereas values like Plant Senescence Reflectance Index (PSRI) and near-infrared reflectance (NIR) were higher in infected plants.
- The integration of minimal datasets combining the most effective features allowed accurate early identification of infected plants, enhancing the potential for high-throughput crop monitoring in vertical farms.

Reference: Agarwal, A., de Jesus Colwell, F., Bello Rodriguez, J. *et al.* Monitoring root rot in flat-leaf parsley via machine vision by unsupervised multivariate analysis of morphometric and spectral parameters. *Eur J Plant Pathol* **2024**, 169, 359–377. DOI: 10.1007/s10658-024-02834-z

Schematic layout of the experimental vertical farming setup with "deep water culture" hydroponics. Empty slots in the tray were used for cables connected to a submersible air pump (for root aeration) and pipes with circulating water baths to maintain water temperature. Source: Agarwal *et al.*, 2024

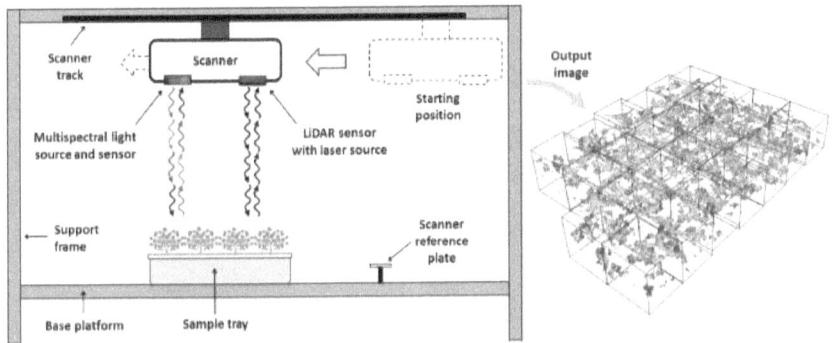

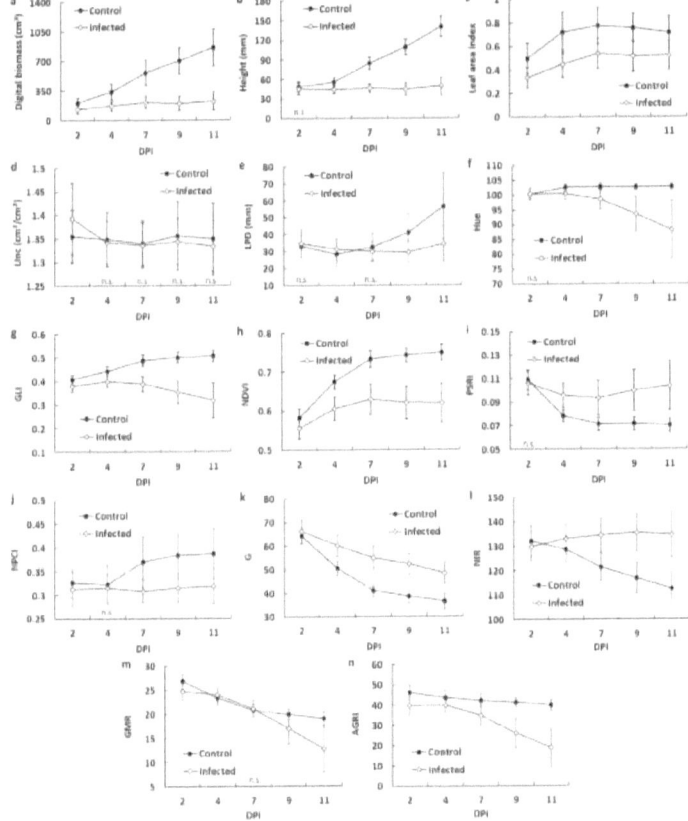

Schematic layout of the imaging setup for 3D-cum-multispectral scanning (left) and a sample output point-cloud image (right). The imaging setup consisted of an overhead scanner equipped with red, green, blue, and near-infrared LEDs along with corresponding sensors for multispectral imaging, as well as a LiDAR sensor for 3D imaging. The scanner moved along a horizontal track from one end of the platform to the other while scanning the samples, and the reference plate assisted in spectral and positional calibration. Boxes shown in the point-cloud image (right) correspond to individual samples (~ 10,000 data points) with a false-colour scheme being used to depict the data from one spectral channel. Source: Agarwal *et al.*, 2024

Digital biomass (a), plant height (b), leaf area index (c), leaf inclination (LInc; d), light penetration depth (LPD; e), Hue (f), Green Leaf Index (GLI; g), Normalised Difference

Vegetation Index (NDVI; h), Plant Senescence Reflectance Index (PSRI; i), Normalised Pigment Chlorophyll ratio Index (NPCI; j), green reflectance (G; k), near-infrared reflectance (NIR; l), green-minus-red reflectance (GMR; m), and Augmented Green-Red Index (AGRI; n) of flat-leaf parsley infected with P. irregulare (main trial). Values have been expressed as mean ± SD (Control: n = 52; Infected: n = 52 for 2–7 DPI, n = 48 for 9 and 11 DPI). "n.s." indicates no statistically significant difference (p > 0.05) between the mean values of control and infected samples at the specified interval (DPI) following one-way ANOVA.
Source: Agarwal et al., 2024

Scientific Report

Integrating Neural Network for Pest Detection in Controlled Environment Vertical Farm

Country: India
Publication Date: 19 May 2022
Main focus: The study presents an integrated system for creating and maintaining an optimal controlled environment for vertical farming, focusing on the use of artificial light and a CNN model-based method for pest detection.

Key findings:

- The CNN model for leaf disease detection achieved a training accuracy of 84.8% and a validation accuracy of 67.2% at the end of the fifth epoch.
- The requirement for artificial light exposure in tomato and chilli plants was found to be approximately 1.2 to 1.6 times higher with 200-watt bulbs compared to 100-watt bulbs, resulting in faster germination, vegetation, and flowering.
- Additional insights from the study include:
- The study highlighted that both tomato and chilli plants required less time for germination, vegetation, and flowering under artificial light compared to natural sunlight. This is attributed to the consistent and controllable intensity of artificial light, unlike the variable intensity of sunlight.
- A CMOS image sensor module was used to capture images of the plants at various growth stages, which were then analysed using deep learning models for pest and disease detection.
- Various image augmentation techniques such as rotation, brightness adjustment, shearing, and zooming were applied to enhance the dataset for training the CNN model.

Reference: Chakraborty A, Das S, Mondal B. Integrating Neural Network for Pest Detection in Controlled Environment Vertical Farm. Indian Journal of Science and Technology **2022**, 15(17): 829-838. DOI: 10.17485/IJST/v15i17.353

Resource Use and Optimization

Use of water, energy, and nutrients can be optimised by adjusting irrigation and lighting based on real-time data and forecasts.

Sensors are needed in vertical farming to optimise the use of water, energy, and nutrients by providing real-time data and forecasts for adjusting irrigation and lighting.

This conserves resources and lowers operational costs and environmental impact.

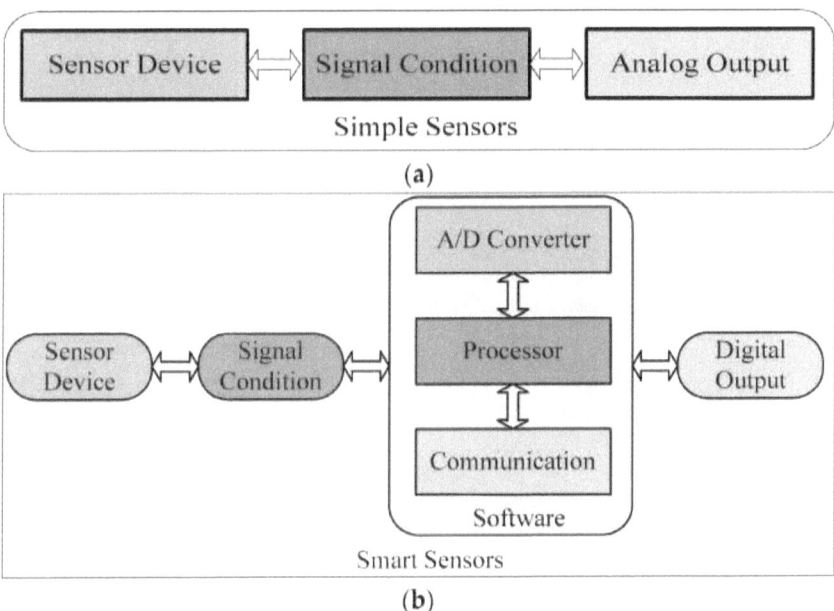

Block diagram of the simple sensors (a) and smart sensors (b). Source: Saad et al., 2021

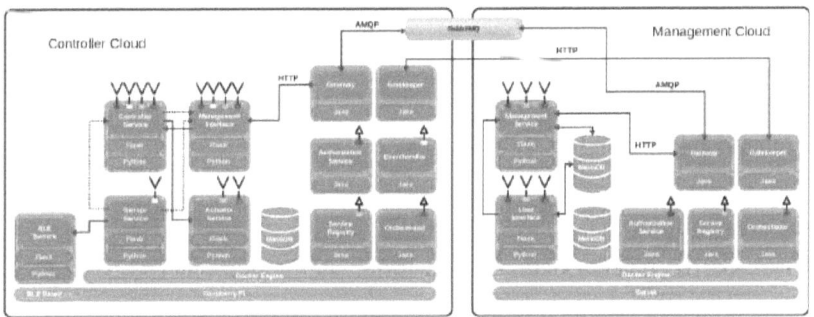

One of the suggested software stacks for indoor vertical farms. Source: Isakovic et al., 2019

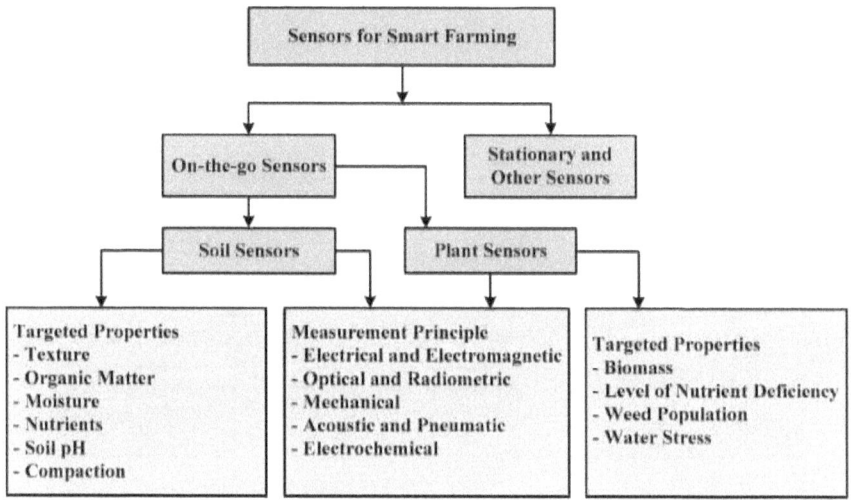

Soil and plant sensors used in smart framing. Source: Saad et al., 2021

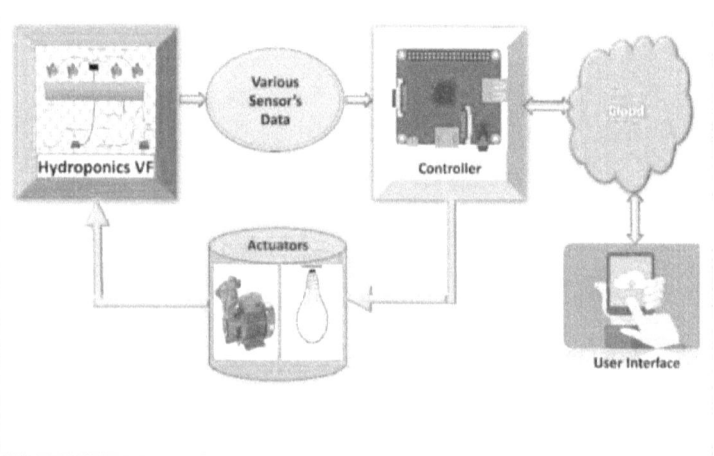

Block diagram of automation in VF with IoT and machine learning algorithms.
Source: Rathor et al., 2024

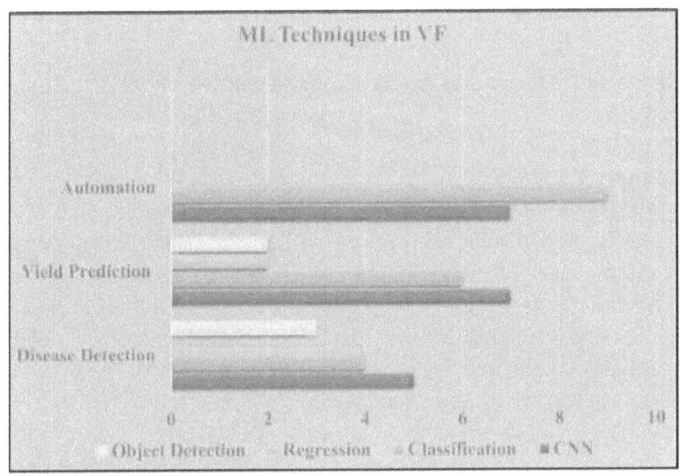

Machine learning techniques in the VF system. Source: Rathor *et al.*, 2024

Modules of an IoT Based Drip Irrigation System for Vertical Farming in Rain Shelter. Source: Sharma *et al.*, 2023

Scientific Report

Realising the Environmental Potential of Vertical Farming Systems through Advances in Plant Photobiology
Country: Switzerland 🇨🇭
Publication Date: 16 June 2022
Main focus: This review explores how advances in plant photobiology can enhance the environmental and commercial efficiency of Vertical Farming Systems

(VFS), which are energy-intensive but offer potential benefits in sustainable food production.

Key findings:
- VFS have high energy demands, primarily for lighting and climate control, leading to significant carbon emissions. The study emphasizes the importance of using renewable energy sources to reduce the environmental footprint.
- The review highlights the role of plant photobiology in optimizing light spectra, intensity, and photoperiod to enhance crop yield and quality while managing energy use efficiently
- VFS can integrate with renewable energy grids, using dynamic lighting strategies to align energy consumption with availability, potentially reducing costs and emissions.
- Bespoke lighting regimes tailored to specific crops can improve yield, quality, and energy efficiency. Understanding plant responses to different light qualities and intensities is crucial for optimizing VFS operations.
- The study suggests that future advancements, including selective breeding and machine learning, could further improve the efficiency and sustainability of VFS.

Resource: de Carbonnel, M.; Stormonth-Darling, J.M.; Liu, W.; Kuziak, D.; Jones, M.A. Realising the Environmental Potential of Vertical Farming Systems through Advances in Plant Photobiology. *Biology* **2022**, *11*, 922. DOI: 10.3390/biology11060922

Controlled environment agriculture enables precise control and monitoring of the growth environments. Light properties that can be varied include photoperiod (daylength), intensity, and quality. These factors interact with other variables (including temperature, humidity, and CO_2 concentration), which are not considered in this review. Future progress in this field will be accelerated by the improved measurement of photosynthetic performance, crop quality

Scientific Report

Vision Based Modeling of Plants Phenotyping in Vertical Farming under Artificial Lighting
Country: Italy
Publication Date: 10 October 2019
Main focus: The paper presents a novel method for vision-based phenotyping of plants in indoor vertical farming, focusing on 3D modelling and deep segmentation of plant leaves under artificial lighting to predict plant growth characteristics.

Key findings:
- The study developed a method combining 3D modelling and deep learning to segment leaves and predict plant growth parameters, including height, weight, and leaf area, using vision-based techniques.
- The research involved extensive data collection, including 2,592 manual data points and 1,728 images, to validate the accuracy of vision-based predictions against traditional manual measurements.
- The method provides a means to optimise environmental controls and resource use in vertical farming, potentially enhancing crop yield and quality without the need for human intervention.

Reference: Franchetti, B.; Ntouskos, V.; Giuliani, P.; Herman, T.; Barnes, L.; Pirri, F. Vision Based Modeling of Plants Phenotyping in Vertical Farming under Artificial Lighting. *Sensors* **2019**, *19*, 4378. DOI: 10.3390/s19204378

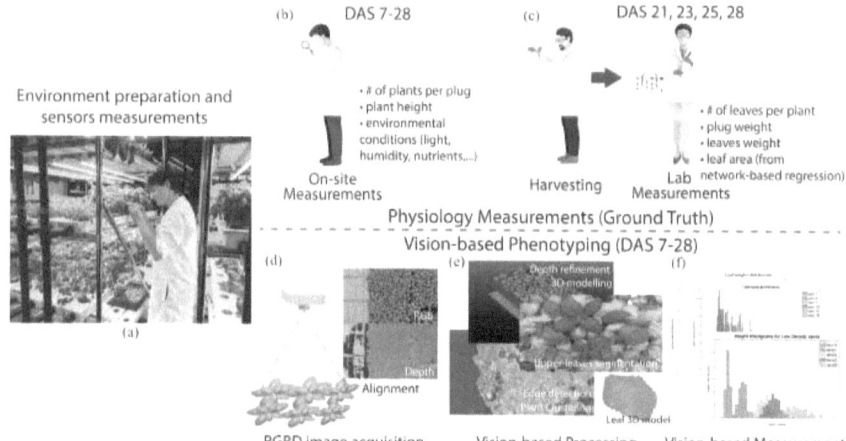

The figure shows the schema of the two parallel computational processes leading to automatic measurements: the experiment setting (a); the manual data collection via on-site (b) and lab measurements (c); the images acquisition and alignment (d); and the ML and CV methods (e) to predict phenotype features, such as plants height, leaf area, and leaf weight (f). Source: Franchetti *et al.*, 2019

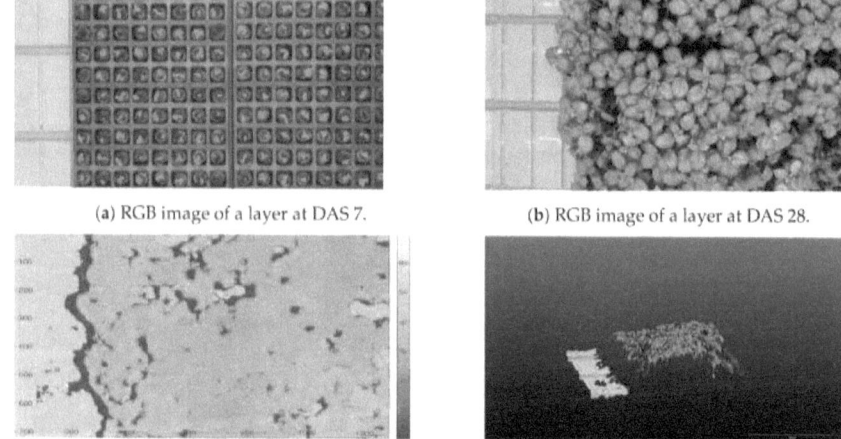

Examples of data collected for assessment include 1,728 colour images and 1,728 depth maps of the plants. Source: Franchetti *et al.*, 2019

Scientific Report

State of the Art of Urban Smart Vertical Farming Automation System: Advanced Topologies, Issues and Recommendations
Country: Malaysia
Publication Date: 13 June 2021
Main focus: The review covers the advancements in automation technologies for urban smart vertical farming (USVF), highlighting topologies, challenges, and recommendations for improving the efficiency and scalability of these systems.

Key findings: The USVF market is projected to grow significantly, with estimates suggesting an increase from $4.4 billion in 2019 to $15.7 billion by 2025. The study emphasises the importance of integrating Internet of Things (IoT) technologies and automation to reduce human intervention, improve crop yield, and manage resources more effectively.

Reference: Saad, M.H.M.; Hamdan, N.M.; Sarker, M.R. State of the Art of Urban Smart Vertical Farming Automation System: Advanced Topologies, Issues and Recommendations. *Electronics* **2021**, *10*, 1422. DOI: 10.3390/electronics10121422

Automation

In large-scale vertical farms with a floor area of 5,000 m² or more, **automation** plays a crucial role in processes like seedling, transplanting, packaging, and internal transportation [Vatistas *et al.*, 2022]. These farms often use autonomous systems, including elevators equipped with irrigation capabilities and cameras for crop inspection. While these technologies help reduce labour costs, they also lead to increased energy demands.

Automation in vertical farming involves **using technology to control and manage farming operations with minimal human intervention**.

> Sensors, irrigation systems, climate control systems, and robotic arms are often **integrated together in an automated vertical farming system** to optimise growth conditions and streamline operations

Scientific Report

Containerized Vertical Farming Using Cobots
Country: United States
Publication Date: 23 October 2023
Main focus: This paper investigates the use of collaborative robots (cobots) to automate key farming operations such as transplantation and harvesting within containerized vertical farms (CVFs), utilising hydroponic systems in mobile shipping containers.

Key findings:
- Cobots demonstrated an overall success rate of 83.8% in automating transplantation and harvesting tasks.
- The transplantation task achieved a success rate of 86.7%, with the cobots accurately inserting saplings into planting slots with a success rate of 80% in challenging conditions.
- Harvesting tasks had a success rate of 80%, with the cobots successfully extracting grown plants from slots despite occlusions caused by plant foliage.
- These results indicate the feasibility and efficiency of using cobots to automate labour-intensive operations in CVFs, enhancing productivity and reducing manual labour.

Reference: Mahalingam D, Patankar A, Phi K, Chakraborty N, McGann R, Ramakrishnan IV. Containerized Vertical Farming Using Cobots. *arXiv.* **2023**. Available from: DOI: 10.48550/arXiv.2310.15385

Vertical Farm in a shipping container with vertical grow panels stacked horizontally.
Source: Mahalingam *et al.*, 2023

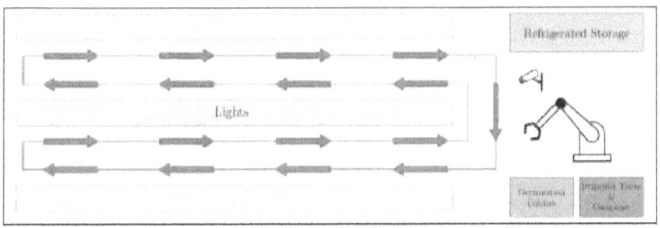

Schematic sketch of our CVF with grow panels moving on conveyors, cobot workstation, and a remotely monitoring farmer. Source: Mahalingam et al., 2023

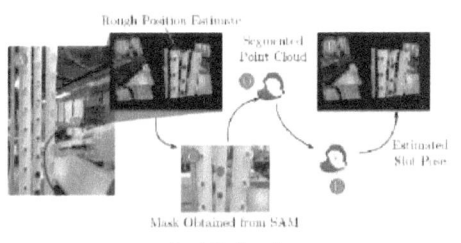

Solution Approach Overview: Left Image - A) Eye-in-hand setup used to capture RGBD images B) Obtained RGBD image from sensor with the rough slot position estimate shown C) Position estimate of slot in the pixel space show as a blue dot and corresponding mask obtained from SAM shown in orange D) 3D points corresponding to slot segmented out of the RGBD image E) Bounding box fit to the 3D points corresponding to the slot for pose estimation F) Pose estimate of slot; Right Image - Schematic sketch of motion estimation from demonstration for a transplanting task G) Recorded demonstration show with * used to represent SE(3) poses to help reduce clutter H) The recorded demonstration is segmented into a sequence of constant screws for identifying the sequence of motion subgroup constraints on the end-effector motion relative to the task-related objects that lie within the region-of-interest of the plant sapling and planting slot I) Transferring the extracted constraints to a new planting slot and determining the final end-effector motion as a sequence of constant screws.Source: Mahalingam et al., 2023

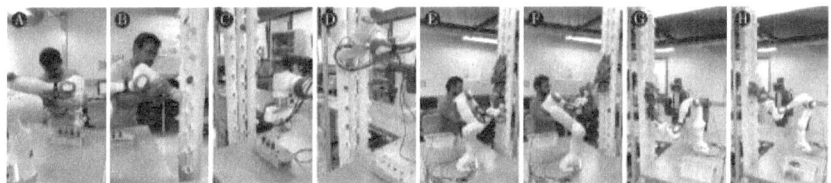

Experimental Setup: A, B: Demonstration for transplanting task, C, D: Execution from provided demonstration for transplanting task, E, F: Demonstration for harvesting task, G, H: Execution from provided demonstration for harvesting task. Source: Mahalingam et al., 2023

A **vertical conveyor system** in vertical farming is a multi-level crop transport solution designed to move plants, trays, or containers between different levels of a multi-story growing facility.

This system typically consists of **a series of conveyor belts, lifts, or automated platforms** that can transport items vertically and horizontally throughout the farm.

It's important because it significantly **enhances operational efficiency** by reducing manual labour, minimising the risk of plant damage during transport, and allowing for precise movement of crops between different growth stages or environmental zones.

This automation enables farmers to **manage larger-scale operations** more effectively, **optimise space utilisation**, and **maintain consistent care for plants** across all levels of the vertical farm. Additionally, it can **integrate with other automated systems** for planting, harvesting, and monitoring, contributing to a more streamlined and productive farming process.

Scientific Report

Design and Validation of an Open-Sourced Automation System for Vertical Farming
Country: United Kingdom
Publication Date: 5 December 2023
Main focus: The study focuses on the development and validation of the Modular Automated Crop Array Online System (MACARONS), an open-sourced platform for plant care, monitoring, and transportation in vertical farming.

Key findings:
- The MACARONS system automates plant care tasks, performing them 30% faster than manual operations.
- The system supports up to 12 payloads and can handle weights of up to 5 kg, with a maximum footprint of 750 mm x 500 mm, making it scalable for various farm sizes.
- The system can be constructed for approximately GBP 2,241.72 (USD 2,793.82), providing a cost-effective solution for automation in vertical farming.
- The platform is fully open-sourced, allowing for customization and upgrades, with software written in Python and MicroPython.

- Automation of plant transport, monitoring, and care can significantly reduce labor costs, which are typically a major expense in vertical farming operations.

Reference: Wichitwechkarn, V., Rohde, W., & Choudhary, R. (2023). Design and validation of an open-sourced automation system for vertical farming. HardwareX, 16, e00497.<u>DOI: 10.1016/j.ohx.2023.e00497</u>

CAD overview image of MACARONS2. Each component is separated by colour for clarity. The orange components include the mover (left), elevator-carriage (back) and carriage (right). The grey components include the shelves (left and right) and elevator (back). For clarity only one carriage is included. Source: Wichitwechkarn *et al.*, 2023

Image of the electric winch (left) and the wire management (middle and right). The red box highlights where the electric winch is mounted and the red arrows show the wire management using V-slot covers. Source: Wichitwechkarn *et al.*, 2023

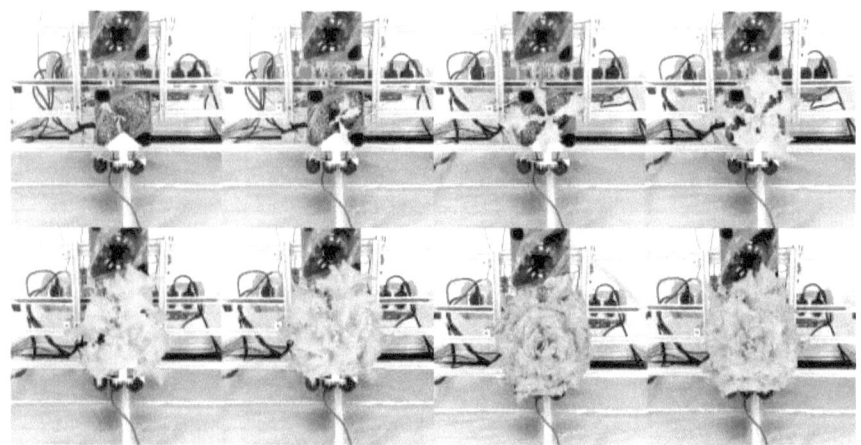

Example image data of a single lettuce plant. The images are taken in order from top left to the bottom right over four weeks. Source: Wichitwechkarn *et al.*, 2023

Scientific Report

Design and Validation of an Automated Storage and Retrieval System for Vertical Farming
Country: South Korea
Publication Date: 15 November 2023
Main focus: The study introduces and validates a novel automated storage and retrieval system (AS/RS) tailored for vertical farming, aimed at optimising space utilisation and operational efficiency.

Key findings:
- The proposed AS/RS model improved time efficiency by 37.54% to 73.22% depending on the logistics quantity, compared to conventional models.
- Vertical farming using the AS/RS system demonstrated potential yield improvements of up to 6,000 times per hectare for certain crops, compared to traditional farming methods.
- The stackable bidirectional infinite-loop module significantly enhanced spatial utilisation, supporting both horizontal and vertical expansion.
- The system addresses labour shortages by reducing the need for manual labour, with enhanced efficiency in logistics movement.
- The AS/RS system in vertical farms promotes resource-efficient crop growth, minimising water and land usage while ensuring consistent year-round production.

Reference: Min, K.; Lim, D. Designing Automated Logistics Warehouse Stackable Bidirectional Infinite-Loop Modules. *Appl. Sci.* **2023**, *13*, 12472. DOI: 10.3390/app132212472

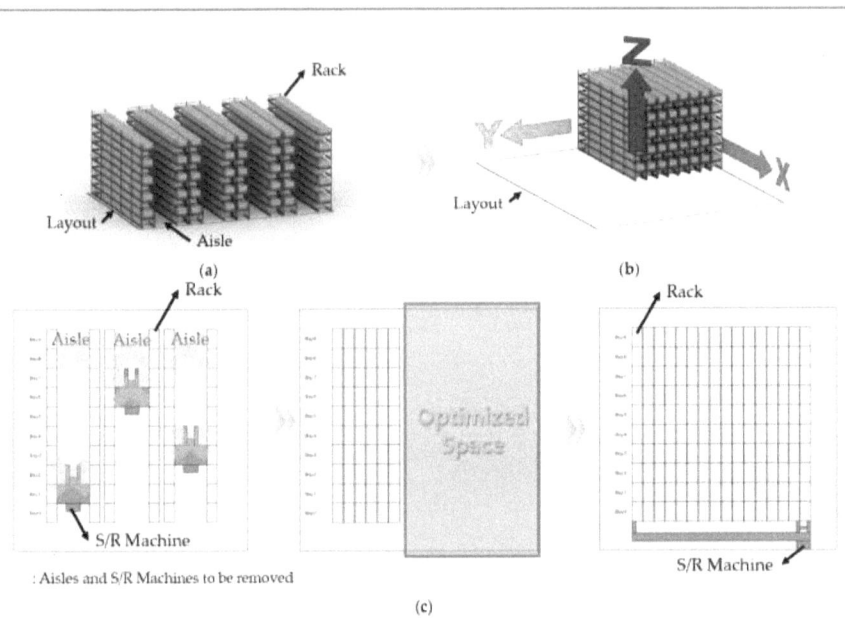

(a) General form of logistics warehouse; (b) structure of a logistics automated warehouse with stackable bidirectional continuous track modules; (c) advantages in space and area due to secured space. Source: Min et al., 2023

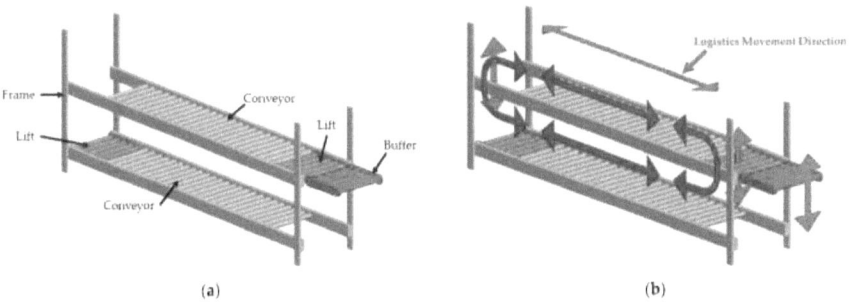

a) Concept diagram of bidirectional continuous track modules; (b) logistics movement direction of bidirectional continuous track modules where the red arrows represent the direction of logistics movement.
Source: Min et al., 2023

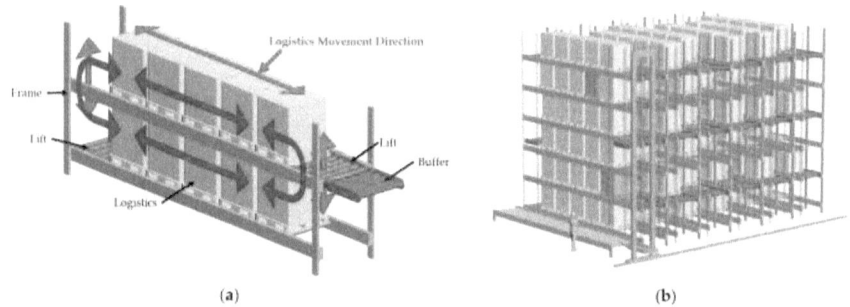

(a) Concept diagram of the pre-transfer system applied in the research; (b) concept diagram of the constructed rack structure. Source: Min *et al.*, 2023

4. Vertical Farming Economics

Establishing a vertical farm requires **significant initial investment** and **ongoing operational costs**. The startup phase involves substantial **expenses for facility acquisition or construction**, as well as **installation of necessary systems**.

> Initial investments can vary widely, from **$100,000-$300,000 for small-scale operations** to **over $10 million for large, high-tech facilities**.

Operating costs present ongoing challenges. Major expenses include:

1. Energy: Typically 25-30% of operating costs
2. Labour: Skilled technicians, workers, and management staff
3. Supplies: Seeds, nutrients, packaging materials
4. Maintenance and equipment replacement
5. Rent or mortgage payments
6. Insurance, marketing, distribution, and regulatory compliance

Operating costs often account for **60-80% of revenue**, necessitating careful management and optimization.

Scientific Report

What You May Not Realize about Vertical Farming
Country: United States
Publication Date: 11 April 2022
Main focus: This paper explores the various challenges and considerations in planning and operating a vertical farm, aiming to provide industry-specific insights for investors and growers.

Key findings:
- Vertical farming can reduce water consumption by up to 90% compared to traditional farming methods.
- Significant investments have been made in vertical farming, with over USD 1 billion invested globally in the past decade.
- Energy costs are a major concern, as they can constitute up to 30% of the operating expenses due to the reliance on artificial lighting and climate control systems.

- Vertical farming is likely to remain focused on high-value crops such as leafy greens, herbs, and some fruiting crops like strawberries and tomatoes, due to higher production costs.
- The carbon footprint of vertical farming can be significantly reduced by exploiting renewable energy sources for lighting and climate control.
- Collaboration and data sharing among vertical farmers and researchers are essential for advancing the industry and overcoming the tendency to "reinvent the wheel" with each new project.

Reference: Lubna, F.A.; Lewus, D.C.; Shelford, T.J.; Both, A.-J. What You May Not Realize about Vertical Farming. *Horticulturae* **2022**, *8*, 322. DOI: 10.3390/horticulturae8040322

Despite high costs, vertical farming offers **potential benefits such as higher yields per square foot** and **year-round production**.

Factors that can impact profitability include the following aspects:

→ Larger vertical farming setups can **spread fixed costs over more units**, reducing the average cost per unit produced.
→ The **choice of location** can significantly influence costs related to energy consumption, labour availability and rates, and real estate prices, all of which impact overall profitability.
→ Selecting **high-value or high-demand crops** can maximise revenue and profit margins, while also considering the growth cycle and yield of different crops.
→ The **extent of automation and advanced technologies** used can reduce labour costs and increase production efficiency, though it may involve high initial investments.
→ Implementing **energy-efficient practices and technologies** can lower operational costs, especially in energy-intensive processes like lighting and climate control, thus enhancing profitability.

Economic efficiency of applying air anions to the cultivation of two cultivars of lettuce in a plant factory. Source: Lee *et al.*, 2022

Type of Lettuce	Net Profit for Unit Area (15 m2)	Net Profit for 1500 m2
Red leaf	USD 577.5	USD 57,750
Lollo bionda	USD 696.5	USD 69,650

Cultivation period: 1 year (7 harvest); cultivation area: 15 m2 × 4 tiers × 25 lines = 1500 m2; lettuce price: USD 1.6 and 1.7 per 100 g for red leaf and Lollo bionda lettuce, respectively; exchange rate applied as of 5 July 2022 (KRW 1300/USD 1).

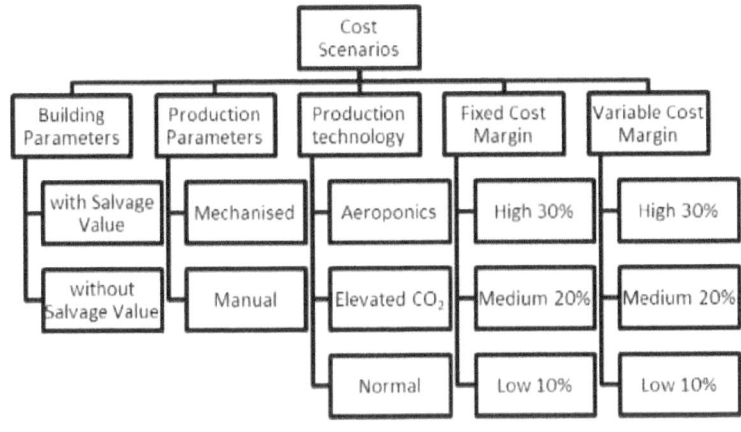

Identified Cost Scenarios. Source: Banerjee & Adenaeuer, 2014

Scientific Report

Life Cycle Cost Analysis of Rooftop Gardens Using OpenLCA
Country: India

Publication Date: 2022

Main focus: The study conducts a Life Cycle Cost Analysis (LCCA) of rooftop gardens compared to conventional roofs in commercial buildings, focusing on assessing economic viability and sustainability.

Key findings: The initial cost of installing a rooftop garden was found to be 5.2 times higher than a conventional roof, but the overall life cycle cost was 5.25% lower due to energy savings and reduced maintenance costs. The rooftop garden led to significant energy savings, reducing the life cycle cost to INR 4085 per sq. ft., compared to INR 4310 per sq. ft. for the conventional roof.

Reference: Barathi, R. D., & Vidjeapriya, R. Life Cycle Cost Analysis of Rooftop Gardens Using OpenLCA. IOP *Conference Series: Earth and Environmental Science* **2022**, 1086(1), 012006. DOI:10.1088/1755-1315/1086/1/012006

Costs considered in the evaluation of LCC.
Source: Barathi & Vidjeapriya, 2022

List of costs	Conventional Roof Amount (INR per sq. ft.)	Rooftop garden Amount (INR per sq. ft.)
Initial costs	155	800
Maintenance costs	300	600
Renovation costs	175	-
Energy costs	3680	2685

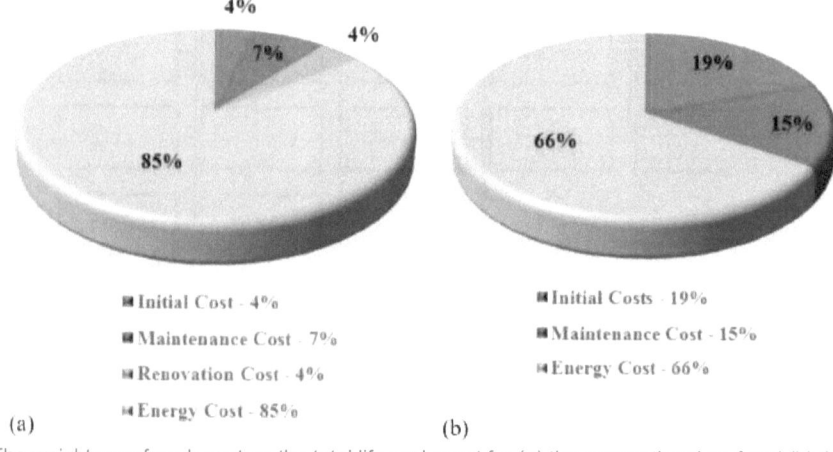

The weightage of each cost on the total life cycle cost for (a) the conventional roof and (b) the rooftop garden. Source: Barathi & Vidjeapriya, 2022

Scientific Report

Economics of Vertical Farming: Quantitative Decision Model and a Case Study for Different Markets in the USA
Country: United States
Publication Date: September 29, 2021
Main focus: The study quantitatively models and evaluates the economic prospects of vertical farming as a business venture in a competitive marketplace under various circumstances, proposing a generalised quantitative framework to evaluate vertical farming compared to traditional farming.

Key findings:
- Vertical farming is projected to grow from $3.16 billion in 2018 to $22.07 billion by 2026.
- Benchmark values for traditional farming versus vertical farming: labour per kg (0.014 manH/kg vs. 0.066 manH/kg), energy consumption per kg (0.575 KWH/kg vs. 5.75 KWH/kg), and water consumption per kg (250 L/kg vs. 20 L/kg).
- The study identifies the best and worst markets for implementing vertical farming in the US by evaluating relative profit and risk across several locations.

Reference: Moghimi F, Asiabanpour B. Economics of Vertical Farming: Quantitative Decision Model and a Case Study for Different Markets in the USA. *Research Square*. **2021**. DOI: 10.21203/rs.3.rs-943119/v1

Benchmark performance values for practising tradition and vertical farming. Source: Moghimi & Asiabanpour, 2021

Parameter	Traditional Agriculture (filed grown)	Vertical Farming
Production per year benchmark (kg)	907184.7	907184.74
Labor per kg (manH/kg)	0.014	0.066666667
Energy consumption per Kg (KWH/kg)	0.575	5.75
Water consumption per kg (L/Kg)	250	20
Yield (Kg/M^2/Year)	4	150.2334586
Land ratio for 1 kg production in a year(M^2/kg)	0.25	0.006656307

Regional rates with their references. Source: Moghimi & Asiabanpour, 2021

City Name	State Initial	Median Farm Labor Cost (USD/Hour)	Industrial Energy Rate (USD/KWH)	Fertile Land Monthly Rent (USD/Acre)	Non-Fertile land rent (USD/Acre)	Insurance Subsidy %	Insurance Premium per Acre (USD/Acre)	Water Price (USD/L)	Price of lettuce(USD/Head)
Austin	TX	11.61	0.054	42.5	6	0.7	74	0.0012	1.99
Boston	MA	17.36	0.139	88.5	35	0.63	160	0.0014	1.79
Chicago	IL	16.6	0.066	224	41	0.57	43	0.0012	1.99
Des Moines	IO	14.6	0.062	230	59	0.54	32	0.001	1.99
Los Angeles	CA	16	0.12	423	13	0.59	31	0.0016	1.79
Miami	FL	12.67	0.072	110	15.5	0.63	53	0.0012	1.99
New York	NY	15.6	0.052	66	26	0.7	26	0.0014	1.49
Source		(US Department of Labor and Statistics, 2020)	(US Energy Information Administration, 2020)	(USDA, 2020)	(USDA, 2020)	(USDA, 2020)	(USDA-NASS, 2020)	(USDA-NASS, 2020)	(US Department of Energy, 2017)	(Wholefoods Market, 2020)

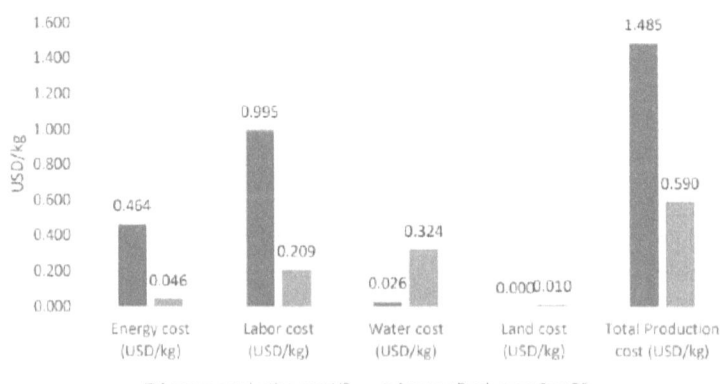

Average production cost per unit for both farming practices.
Source: Moghimi & Asiabanpour, 2021

Scientific Report

Process Simulation and Techno-Economic Analysis of Large-Scale Bioproduction of Sweet Protein Thaumatin II

Country: United States
Publication Date: 12 April 2021
Main focus: This study presents a process simulation and techno-economic analysis of the large-scale production of sweet protein Thaumatin II using various molecular farming platforms, including indoor vertical farming.

Key findings:
- The proposed large-scale production facility aims to produce 50 metric tons of Thaumatin II annually, utilizing transgenic *Nicotiana tabacum* and *Nicotiana benthamiana* plants.
- The capital expenditure (CAPEX) for the field-grown transgenic thaumatin production facility is estimated at $10.2 million, with an annual operating cost (AOC) without depreciation of $3.59 million.
- The cost of goods sold (COGS) without depreciation for the field-grown thaumatin production is $71.7/kg, highlighting the economic feasibility of large-scale bioproduction compared to current methods.

Reference: Kelada, K.D.; Tusé, D.; Gleba, Y.; McDonald, K.A.; Nandi, S. Process Simulation and Techno-Economic Analysis of Large-Scale Bioproduction of Sweet Protein Thaumatin II. *Foods* **2021**, *10*, 838. DOI: 10.3390/foods10040838

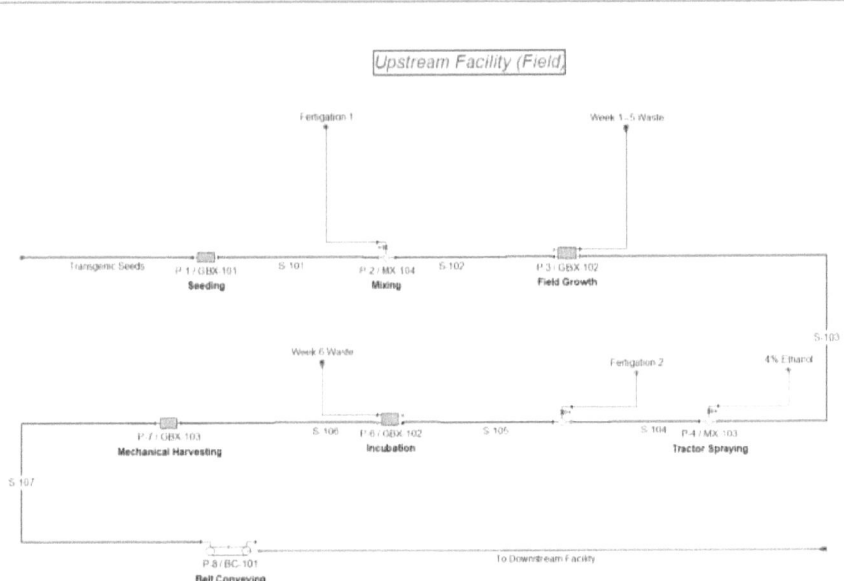

Model flowsheet for base case upstream transgenic production facility.
Source: Kelada *et al.*, 2023

Capital expenditure (CAPEX), annual operating costs (AOC), and cost of goods sold (COGS) for transgenic thaumatin production facilities. Depreciation is based on the 10-year straight line method.
Source: Kelada *et al.*, 2023

Facility	Upstream (Field)	Upstream (Indoor)	Downstream (With Chromatography)
CAPEX ($ million)	10.2	186	115
AOC without depreciation ($ million)	3.59	96.1	25.0
COGS without depreciation ($/kg)	71.7	1920	499
AOC with depreciation ($ million)	4.22	110	35.3
COGS with depreciation ($/kg)	84.5	2200	706
	Upstream (Field)	Upstream (Indoor)	Downstream (Without Chromatography)
CAPEX ($ million)	9.22	157	83.7
AOC without depreciation ($ million)	3.00	80.3	12.8
COGS without depreciation ($/kg)	60.0	1600	258
AOC with depreciation ($ million)	3.56	94.3	20.4
COGS with depreciation ($/kg)	71.2	1890	408

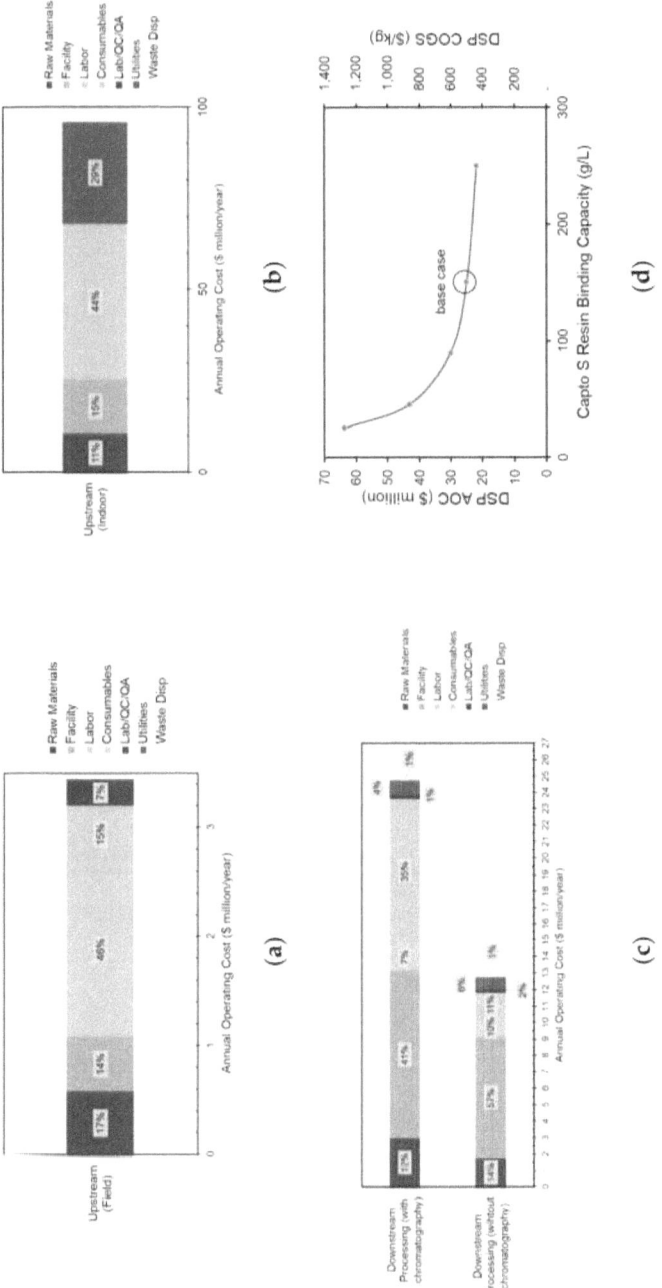

Annual operating costs breakdown per category for (**a**) field upstream transgenic facility, (**b**) indoor upstream transgenic facility, (**c**) base case downstream processing facility and without chromatography unit operation, (**d**) effect of resin binding capacity on DSP AOC and COGS. Depreciation costs are excluded. AOC, annual operating costs; COGS, cost of goods; DSP, downstream processing. Source: Kelada et al., 2023

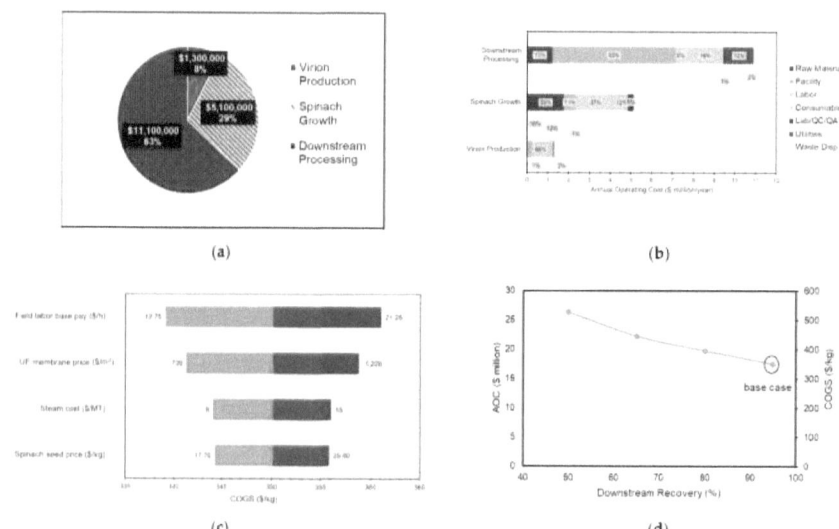

Spinach made annual operating cost breakdown (**a**) per facility section (**b**) section's category. (**c**) Tornado analysis for facility's top cost drivers. (**d**) Effect of varying downstream recovery on AOC assuming a constant target production level of 50 MT/year. AOC, annual operating costs; COGS, cost of goods sold; MT, metric tons. Source: Kelada et al., 2023

Scientific Report

Energy and Cost Analysis for Crop Production in a Vertical Farm

Country: Italy
Publication Date: 1 December 2023
Main focus: This study evaluates the energy consumption and associated costs of crop production in vertical farming systems, focusing on optimizing energy efficiency and economic feasibility across different climate conditions.

Key findings:
- The energy consumption for lettuce production was highest in Riyadh at 10.1 GWh/year, 38% higher than in Naples and 86% higher than in Stockholm.
- The specific cost for energy (SCE) was lowest in Stockholm at €0.85/kg, compared to €1.35/kg in Naples and €1.75/kg in Riyadh.
- Lighting systems accounted for 65-85% of total energy consumption, while HVAC systems contributed 15-20%, highlighting the critical role of efficient lighting in reducing overall energy use.

Reference: Arcasi A, Mauro AW, Napoli G, Tariello F, Vanoli GP. Energy and cost analysis for a crop production in a vertical farm. *Applied Thermal Engineering*. **2024**. 239:122129. DOI: 10.1016/j.applthermaleng.2023.122129

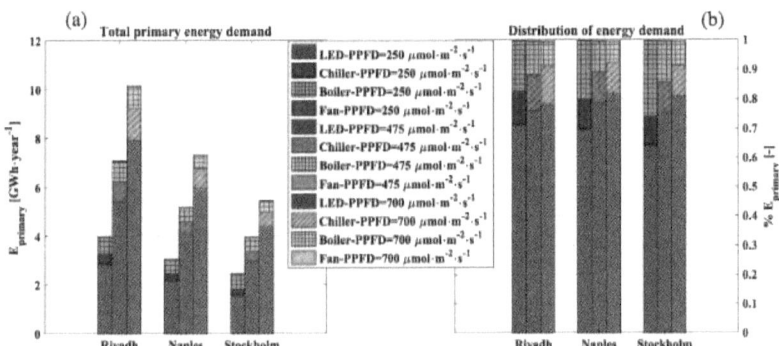

Effect of the PPFD level on the total energy demand (a) dimensional plot and (b) percentage plot for a leaf surface temperature of 24 °C and a LED efficiency of 0.6 for the three considered climate zones of Riyadh, Naples and Stockholm. Source: Arcasi et al., 2023

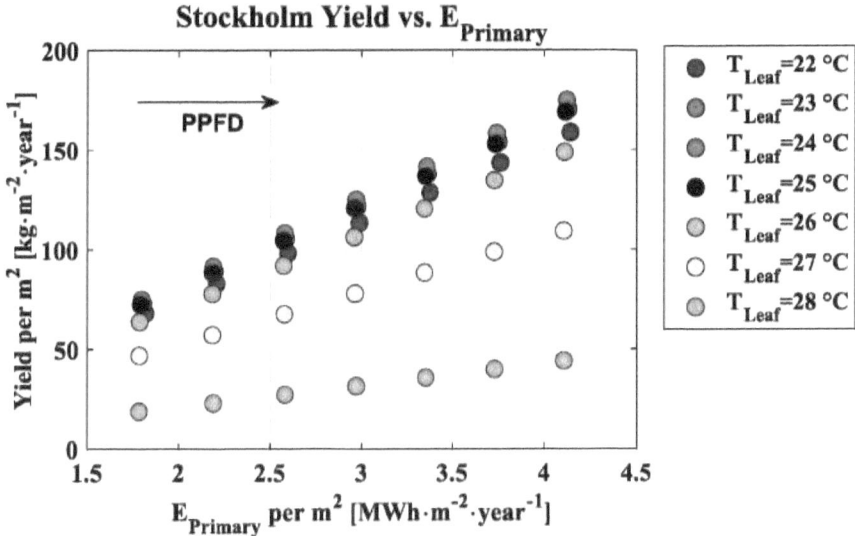

Yield per m2 with respect to primary energy consumption per m2s for a fixed LED efficiency of 0.6 for the climate zone of Stockholm. Source: Arcasi et al., 2023

5. Design and Construction

This chapter, *Design and Construction*, explores the essential elements of creating efficient and functional vertical farming facilities.

Architectural Considerations highlights the unique challenges and opportunities presented by vertical farm design. This section covers aspects such as **space optimization, light penetration, and circulation patterns** within these specialised structures.

Structural and Environmental Design Elements focuses on the **technical aspects of vertical farm construction**. It addresses critical factors such as load-bearing requirements, materials selection, and the integration of complex systems for lighting, irrigation, and climate control.

The final part, *Design and Integration with Existing Buildings*, examines strategies for **incorporating vertical farming systems into pre-existing urban structures**. This section discusses the challenges and benefits of retrofitting buildings for agricultural use, including considerations for structural modifications and utility integration.

Architectural Considerations

Architectural design for vertical farms is a complex process that must balance functionality, efficiency, and often, aesthetic appeal.

Concept of Interior view of Vertical Farm. Source: Cicekli & Barlas, 2014

Space utilization is a primary concern in vertical farm design. Layouts must accommodate **crop production, processing, and logistics** while allowing for efficient workflows. Structural integrity is crucial, as the building must support the weight of growing systems, water, and equipment.

Light management is an important factor. Designs typically incorporate artificial lighting systems and may also optimise natural light use where applicable. **Climate control** is equally significant, often involving sealed environments and HVAC systems to maintain consistent growing conditions

Water and nutrient delivery systems are integral components. Designs usually include plumbing for hydroponic or aeroponic systems, along with water management facilities. Biosecurity measures, such as airlocks and separated zones, are often incorporated to minimise contamination risks.

Energy use is a consideration in vertical farm design. Some designs may include renewable energy systems or passive environmental control features. Automation is frequently a factor, with infrastructure for monitoring and control systems often integrated into the design.

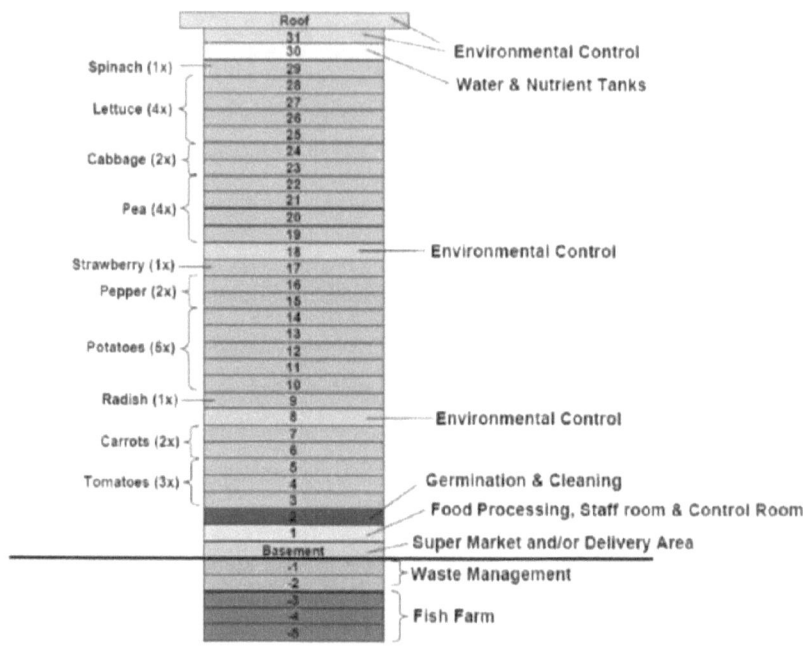

Layout of the Vertical Farm. Designed in a CE Study by the author at DLR Bremen.
Source: Banerjee & Adenaeuer, 2014

The **external appearance** of vertical farms can vary. Designs may take into account the visual impact on surrounding areas and sometimes include public-facing elements such as viewing areas.

Interior designs typically address worker needs, including ventilation, workspace design, and safety systems. Other common elements include waste management facilities, vertical transportation systems, and adaptable growing spaces.

Vertical farm architecture continues to develop as the field evolves. Ongoing changes in technology, sustainability requirements, and urban food production needs may influence future designs in this area towards including wider adoption of **aquaponics** and **livestock management** in vertical farming.

Aquaculture in vertical farming.
Source: Cicekli and Barlas, 2014

Production of livestock in vertical farming.
Source: Cicekli and Barlas, 2014

Farmhouse, vertical farm prototype (2019), designed by Chris Prech.
Source: Bisiani *et al.*, 2022

Structural and Environmental Design Elements

From an architectural and engineering perspective, a vertical farm typically consists of **multi-story structures** that incorporate hydroponic, aeroponic, or aquaponic systems to maximise space efficiency and resource utilisation.

Scientific Report

Fabrication and Performance Evaluation of Vertical Farming Structures
Country: India
Publication Date: 8 May 2021
Main focus: This study focuses on designing, fabricating, and evaluating the performance of vertical farming structures suitable for homestead and urban farming environments.

Key findings:

- Two vertical farming structures, DVFS 1 and DVFS 2, were developed and tested. DVFS 1 showed superior performance, with a 58% yield compared to 42% for DVFS 2.
- Biometric observations indicated that plant height and number of leaves were significantly higher in DVFS 1, with the right side tiers (T1 and T3) performing best due to better light intensity.
- The correlation between plant growth parameters and light intensity was highly significant, with a Pearson correlation coefficient of 0.935 for plant height and 0.86 for the number of leaves.

Reference: Shaheemath Suhara KK, Nair PG. Fabrication and performance evaluation of vertical farming structures. *Journal of Applied and Natural Science.* **2021** Jul 19. DOI: 10.31018/jans.v13iSI.2777

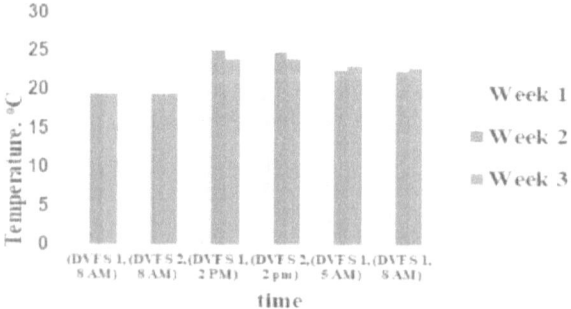

Variations in air temperature in DVFS 1 and DVFS 2 at 8:00 am, 2:00 pm and 5:00 pm.
Source: Shaheemath Suhara & Nair, 2021

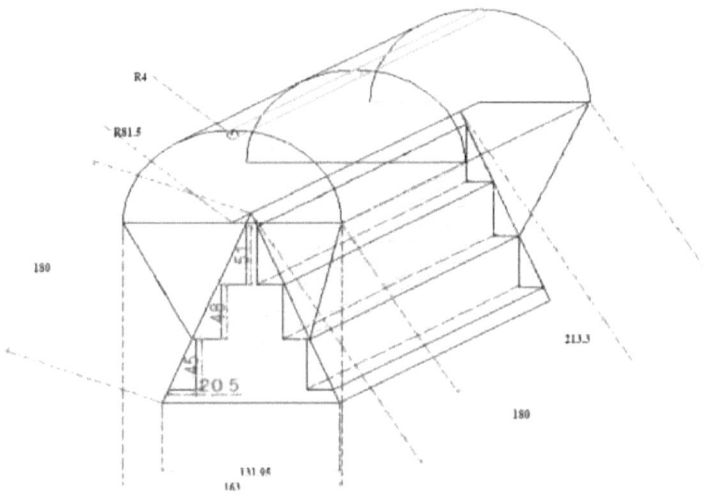

Developed vertical farming structure 1. Source: Shaheemath Suhara & Nair, 2021

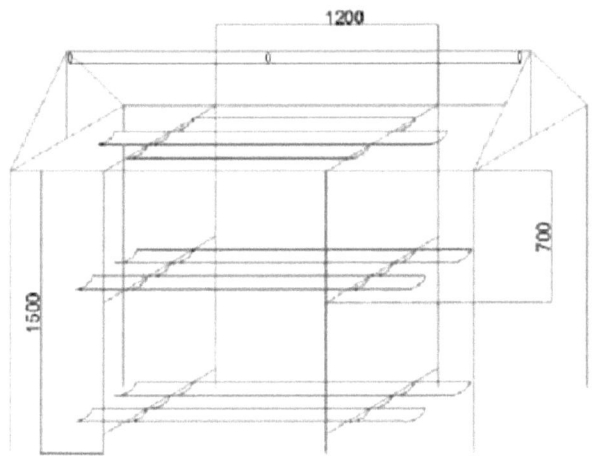

Developed vertical farming structure 2. Source: Shaheemath Suhara & Nair, 2021

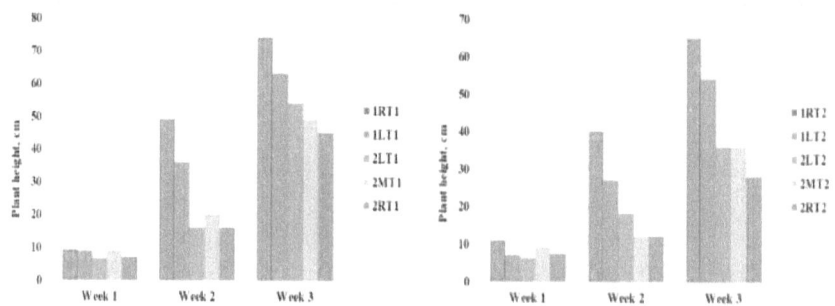

Variation of plant height in T1 (tier 1, on the left side) and T2 (tier 2, on the right side) of DVFS 1 and DVFS 2. Source: Shaheemath Suhara & Nair, 2021

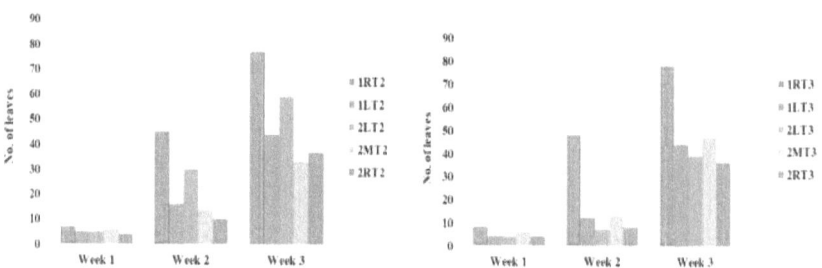

Variation of number of leaves in T2 (tier 2, on the left side) and T3 (tier3, on the right side) of DVFS 1 and DVFS 2. Source: Shaheemath Suhara & Nair, 2021

Thermally well-insulated and airtight walls are essential features in building design of indoor vertical farms, especially for environments requiring precise climate control, such as vertical farms. These walls reduce heat transfer between the interior and exterior, helping to maintain a consistent internal temperature. **Airtight construction** further aids in controlling the internal environment by preventing the entry of outside air and contaminants, which can interfere with climate control systems.

> **Stability of the environment** is important for minimising energy use related to heating and cooling. Insulation helps retain heat during colder periods and blocks excessive heat during warmer times, supporting efficient temperature management.

Proper combination of **insulation and airtightness** is crucial for maintaining the specific conditions necessary for optimal plant growth.

Scientific Report

Computational analysis of the environment in an indoor vertical farming system
Country: USA, Philippines
Publication Date: 31 December 2021
Main focus: This study develops a three-dimensional numerical model to optimize air flow and heat transfer within an indoor vertical farming system, specifically focusing on the effects of turbulence, obstacles, and environmental parameters on plant growth.

Key findings:

- The study found that the most efficient airflow design achieved an objective uniformity of 91.7% due to high spiral flow circulation. The model revealed that certain configurations with low mass flow rates could still provide uniform flow distribution, significantly reducing energy consumption.
- Additionally, horizontal air speeds of 0.3–0.5 m/s enhanced photosynthesis by improving gas exchange between plants and the environment.
- The computational analysis also showed that strategically placing walls and barriers within the vertical farming system can enhance airflow distribution and temperature regulation, thereby optimising plant growth conditions.
- Overall, this model serves as an effective tool for improving the cultivation environment in indoor vertical farming systems.

Reference: Naranjani, B., Najafianashrafi, Z., Pascual, C., Agulto, I., & Chuang, P.-Y. A. Computational analysis of the environment in an indoor vertical farming system. *International Journal of Heat and Mass Transfer* **2022**, Volume 186, 122460. DOI: 10.1016/j.ijheatmasstransfer.2021.122460.

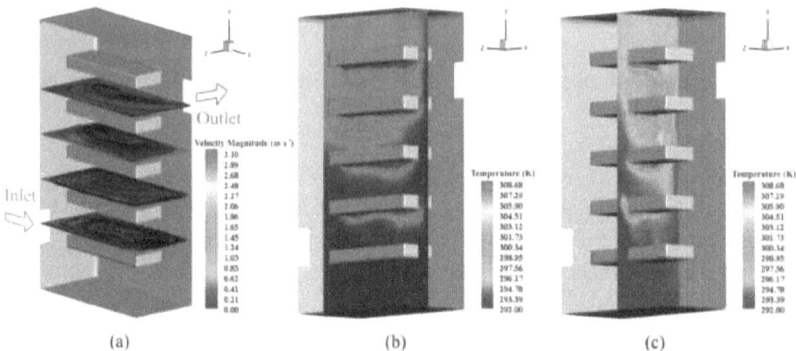

For a scenario where air flows into the system at a rate of 0.3 kilograms per second, the study shows how air moves and its temperature distribution. The velocity contour illustrates the speed of air along the top surface of each plant tray on a horizontal plane. Additionally, temperature contours depict how temperature varies both from front to back (longitudinal direction) and from side to side (transverse direction) of the tray. Source: Naranjani et al., 2022

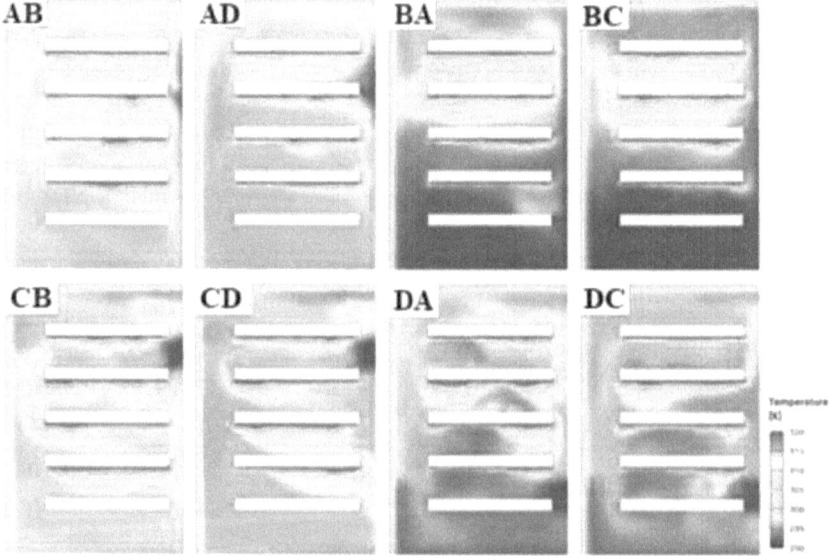

Temperature distribution in two dimensions for eight different inlet-outlet placements with an air flow rate of 0.3 kg per second. Source: Naranjani et al., 2022

Multi-tiered planting structures have become an innovative solution for urban agriculture, allowing us to grow food and plants vertically in limited city spaces. These systems stack plants in multiple layers, turning walls, rooftops, and other vertical surfaces into productive growing areas.

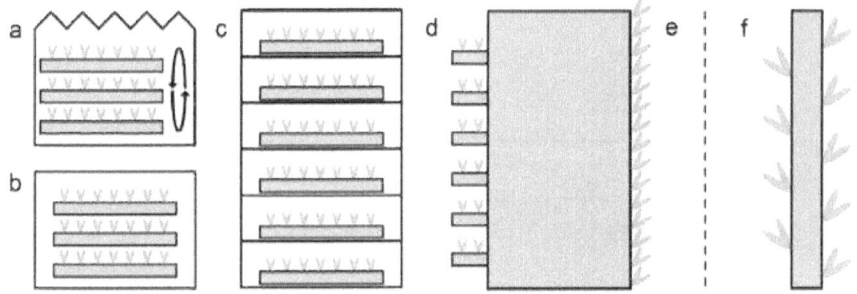

Representation of vertical farming (VF) types.
Stacked horizontal systems comprise multiple levels of horizontal growing surfaces and can be located in glasshouses (a), sometimes with level rotation incorporated, or controlled environment facilities (b). A variation of this approach is that of multi-floor towers (c) where each level is isolated from the surrounding levels. The use of balconies (d) for crop production is another example of VF using stacked horizontal growing surfaces. Vertical growing surfaces include green walls (e), which can be positioned on the side of buildings and other vertical surfaces and cylindrical growth units (f) with vertical arrangements of plants. Source: Beacham *et al.*, 2019

Scientific Report

A New Approach for Vertical Plant Cultivation Maximizing Crop Efficiency
Country: Poland
Publication Date: 17 June 2024
Main focus: This study presents an innovative vertical farming system featuring modular rotating towers designed to optimise crop growth efficiency and resource use.

Key findings:
- The proposed system reduces lighting installation costs by 30% compared to other vertical farming solutions.
- It achieves a 95% reduction in water consumption through a closed ventilation and irrigation circuit.
- The system increases production efficiency with 2240 heads of lettuce per cubic metre annually, significantly outperforming comparable systems like the FreightFarms Greenery S, which produces 595 heads per cubic metre.

Reference: Ptak, M.; Wasieńko, S.; Makuła, P. A New Approach for Vertical Plant Cultivation Maximizing Crop Efficiency. Preprints **2024**, 2024061047. DOI: preprints202406.1047.v1

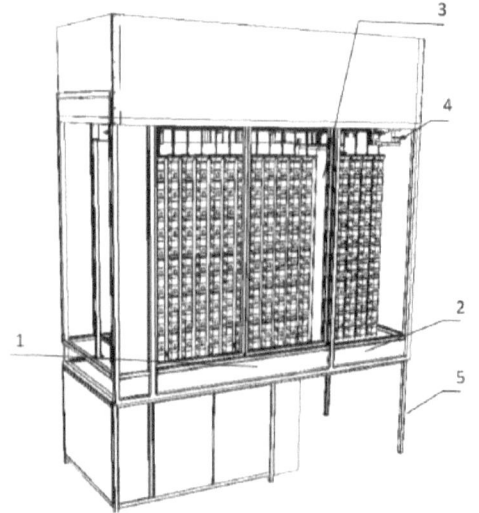

1– climate module "day" A1
2–"night" climate module A2
3–separation device climatic zone divider
4–suspension system
5–tower system frame
6–towers
7–roller for roller shutters for climatic zone partitions
8–exposure panel
9–device sliding system for separating climatic zones
10–tower cart
11–pot
12–mechanism for changing the driving track
13–front door

An axonometric view of a system for tower plant cultivation. Source: Ptak *et al.*, 2024

The vertical farming with the day/night chambers; a) computer-aided design model b) the system during assembly — the towers (in white) are visible c) the prototype module implemented in a grocery shop. Source: Ptak *et al.*, 2024

Testing different types of plants in a module with variable climatic parameters.
Source: Ptak *et al.*, 2024

**Vertical farming cultivation system with the day/night chamber feature.
Source: Ptak *et al.*, 2024**

	Human interaction	Light	Temp.	Humidity	Irrigation	Plant position	Days in the chamber
Loading chamber	Yes	No	Fixed	Fixed	N/A	Horizontally	N/A
Germination chamber	No	No	Fixed	Fixed	Ebb&Flow	Horizontally	3-5
Propagation chamber	No	Yes	Variable	Variable	Ebb&Flow	Horizontally	20-24
A1 – towers' chamber – daytime	No	Yes	Fixed	Fixed	Dripping	Vertically	20-24
A2 – towers' chamber – nighttime	No	No	Fixed	Variable	Dripping	Vertically	
Exit chamber	Yes	Yes	Fixed	Fixed	NFT	Horizontally	2

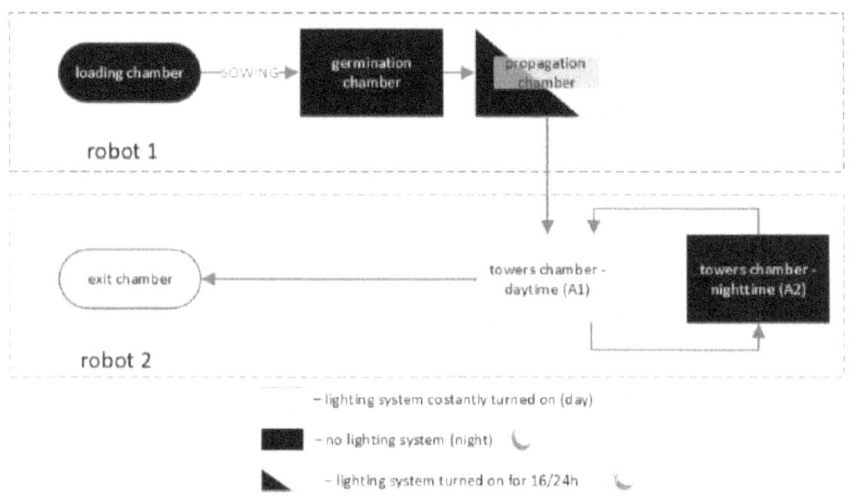

Cultivation process divided into operational zones of robot #1 and #2.
Source: Ptak et al., 2024

There are several types of multi-tiered planting structures, each suited for different environments and crops.

Stacked Containers: Shelves or racks with small planters are ideal for DIY enthusiasts. They are good for growing herbs or small vegetables and can be easily customised and adjusted.

Living Walls: These structures transform plain walls into attractive green displays. Featuring special pockets for plants, they are ideal for growing succulents, small herbs, and ornamental plants, adding greenery to urban areas.

Vertical Hydroponic Systems: These tall structures use water to deliver nutrients directly to plants and require some technological proficiency. They are water-efficient and ideal for growing leafy greens and strawberries.

Trellis Systems: These structures act like climbing gyms for plants, providing frames that support tomatoes, cucumbers, and beans as they grow upwards, maximizing the use of vertical space.

Tiered Raised Beds: These beds look like a staircase for plants, with different levels for planting. They work well on slopes or flat areas and are good for growing a variety of vegetables while being easy to work with.

Green Facade Systems: These systems turn entire buildings into gardens. Using cables or mesh, they support climbing and trailing plants, creating large-scale living walls.

When growing plants on different layers, uneven conditions like temperature and light can **create variations across levels**, leading to inconsistent crop growth. This inconsistency can affect the uniformity and quality of the crops, which is a potential issue for these systems [Beacham *et al.*, 2019]

Choosing the right system depends on factors such as **available space, desired crops, local climate, and structural support**. For example, a sunny wall might be good for a living wall, while a small balcony could benefit from a hydroponic tower. Rooftops can often accommodate a combination of these systems.

Scientific Report

Expanding the Level of Technological Readiness for a Low-Cost Vertical Hydroponic System

Country: Spain
Publication Date: 14 October 2021
Main focus: This paper aims to develop and test a modular, low-cost vertical hydroponic system, assessing its technology readiness level and potential for urban agriculture.

Key findings:
- The prototype system demonstrated high versatility, suitable for various crop types, with a plant density increase of 2-3 times compared to conventional hydroponics.
- LED lighting reduced the vegetative cycle of strawberry plants by 5 days, from 60 to 55 days.
- The system's implementation cost per plant was significantly lower at €19.33 compared to €28.22 for conventional systems, highlighting its economic feasibility.

Reference: Borrero, J.D. Expanding the Level of Technological Readiness for a Low-Cost Vertical Hydroponic System. *Inventions* **2021**, *6*, 68. DOI: 10.3390/inventions6040068

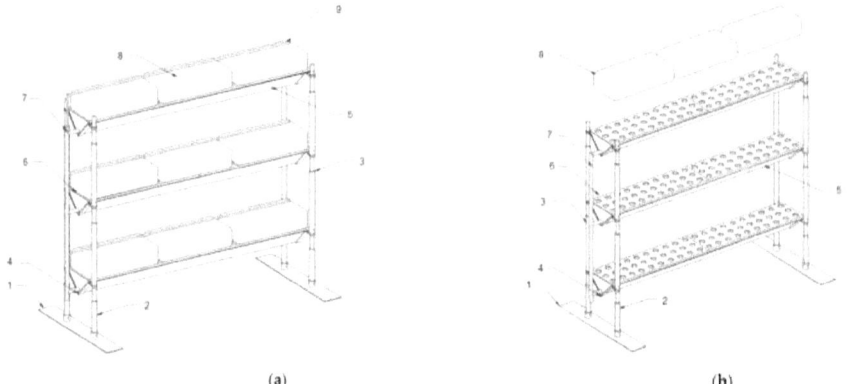

View of the system with all the components that make up the structure. With substrate sacks (**a**) Without substrate sacks (**b**). Source: Borreo, 2021.

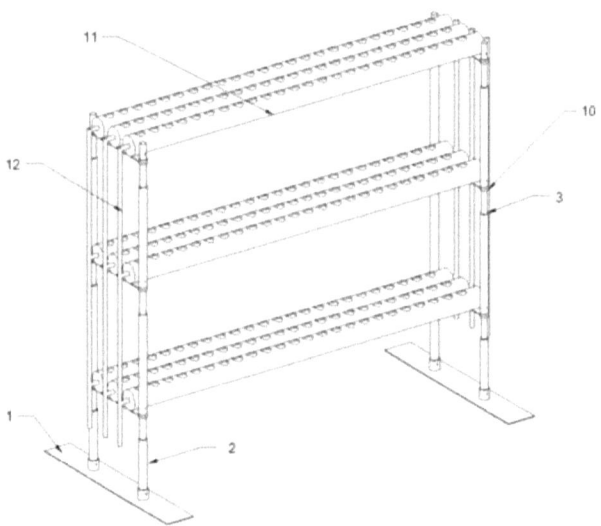

Detail of the adaptation to an aeroponic system. Source: Borreo, 2021.

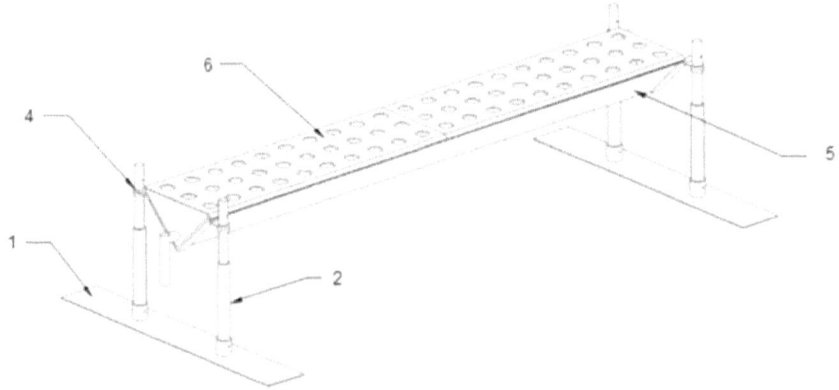

Representation in a free perspective of the adaptation of the system to large crops.
Source: Borreo, 2021

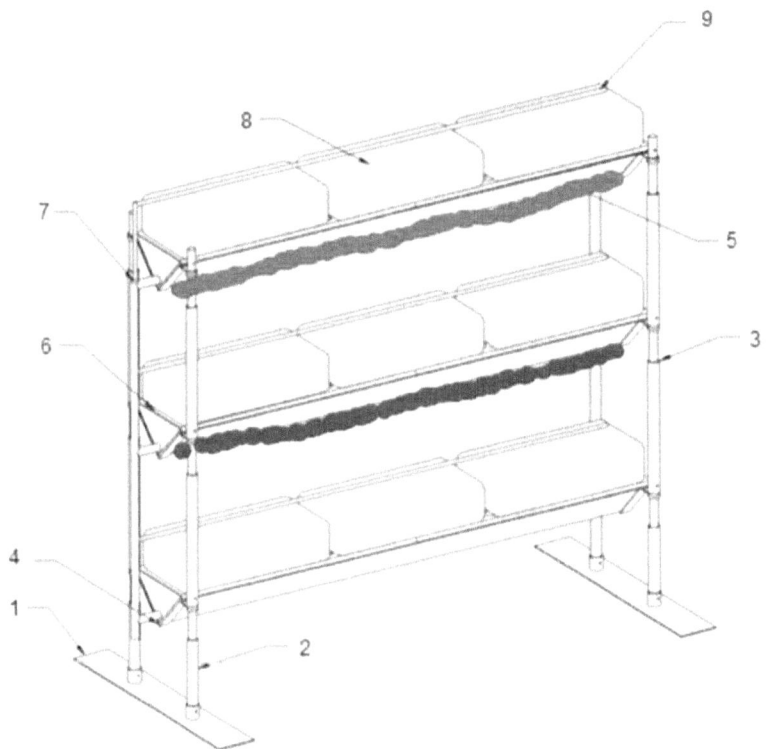

Installed LEDs in the vertical farm system. Source: Borreo, 2021

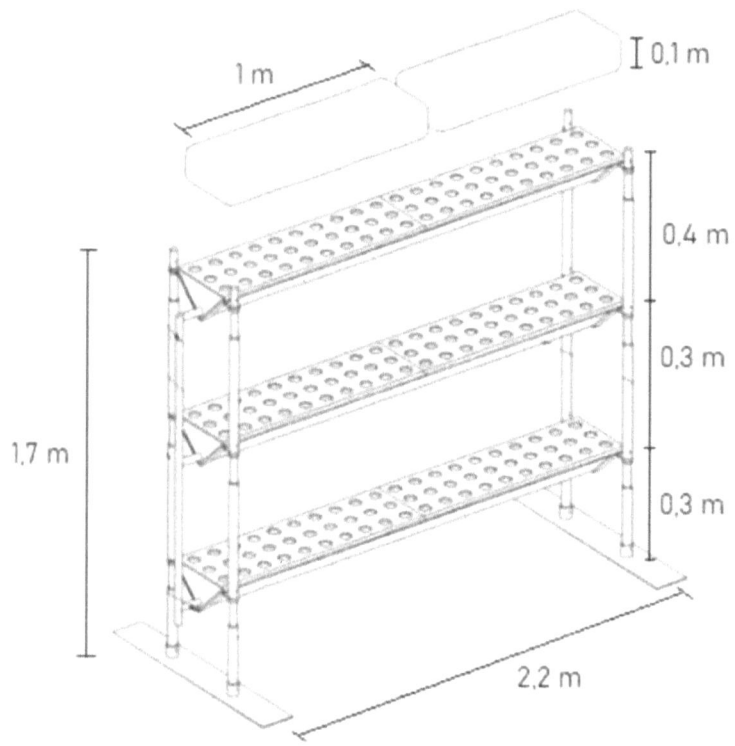

Dimensions of the vertical farm module system. Source: Borreo, 2021

Comparative system costs for a strawberry farm (EUR /ha).
Source: Borreo, 2021

	Conventional Hydroponic Solution	Three-Level Modules
Variable costs	22,680	69,660
Substrate sacks	16,200	48,600
Strawberry plants	6480	21,060
Fixed cost		
Hydroponic system	160,200	337,500
Total implementation cost	182,880	407,160
Total implementation cost per plant	28.22	19.33

Design and Integration with Existing Buildings

The selection of building materials and structures for vertical farms is a critical aspect of their design and construction. These choices not only affect the farm's functionality and efficiency but also its **sustainability and long-term viability**. The design and structure of a vertical garden are influenced by several factors [Mishra et al., 2021]:

1. **Available Materials:** Initially, locally available materials like bamboo or wood / lumber may be used, with the option to upgrade to more durable materials like stainless steel as the farm grows.
2. **Local Preferences**: Crops should be chosen based on suitability to the local environment, tested in trial runs, and switched if they prove unprofitable.
3. **Creativity:** The farm's design should maximize the use of available land and solar radiation.
4. **Crop Management:** Careful planning is required to allocate appropriate space for each crop.
5. **Resources:** Consideration of available energy sources, such as the need for a solar system or other energy supplies.
6. **Space:** The total land available for the farm.
7. **Climate:** Local climate conditions must be taken into account in the design and operation of the garden.

The location **choice between rural and urban areas** is decisive for vertical farming. For example, urban rooftop glasshouses require a significantly higher initial investment, estimated at three times more than conventional ground-based glasshouses, due to necessary building adaptations [Beacham et al., 2019]. The decision between using a glasshouse or controlled environment systems also **impacts costs, particularly regarding artificial lighting and structural requirements**.

Utilizing pre-existing buildings for controlled environment facilities can **lower setup costs** compared to constructing new vertical farming structures.

> Retrofitting existing structures for vertical farming often begins with a **thorough assessment of the building's structural integrity**. Many urban buildings, particularly older warehouses or industrial spaces, have the robust framework necessary to support the weight of growing

systems, water, and equipment. However, **reinforcement is frequently required**, especially when dealing with multi-story setups.

The **façade of the building** is another key consideration. In some cases, existing windows can be utilised to provide natural light for plant growth, potentially reducing energy costs associated with artificial lighting. However, many vertical farming systems require precise light control, leading to the **modification of windows or even their complete coverage** to create a controlled environment. Some innovative designs incorporate **translucent panels or smart glass** that can adjust light transmission, balancing the benefits of natural light with the need for environmental control.

Scientific Report

AGRI|gen: Analysis and Design of a Parametric Modular System for Vertical Urban Agriculture
Country: Palestine
Publication Date: 16 March 2023
Main focus: The study explores the potential of using vertical farming on facades and roofs of buildings in a neighbourhood in Nablus, Palestine, through the development and implementation of parametric tools for analysis and design.

Key findings:
- The neighbourhood's facades can provide about 28,500 m² of farming area, but only half are suitable for daylight-based farming.
- The system can fulfil 350% of local tomato consumption and 237% of cucumber consumption using 25% and 33% of the facades, respectively.
- The modular facade system, consisting of 40,824 units, has an average cost of $55.2/m², totaling $1.7 million for the neighbourhood

Reference: Ghazal, I.; Mansour, R.; Davidová, M. AGRI|gen: Analysis and Design of a Parametric Modular System for Vertical Urban Agriculture. *Sustainability* **2023**, *15*, 5284. DOI: 10.3390/su15065284

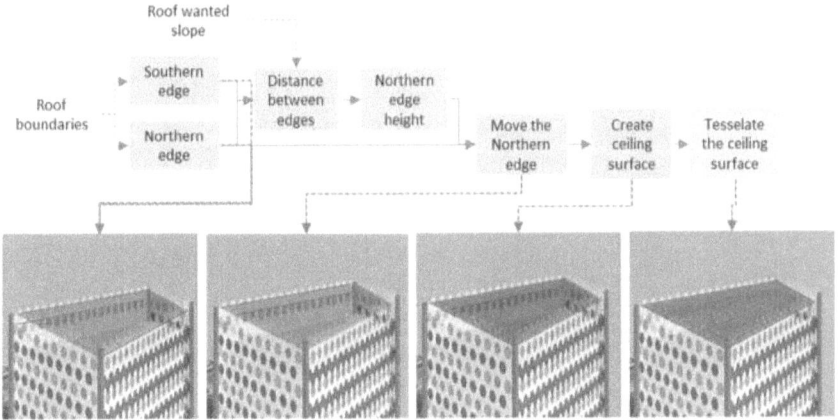

Roof Design algorithm and results. Source: Ghazal et al., 2023.

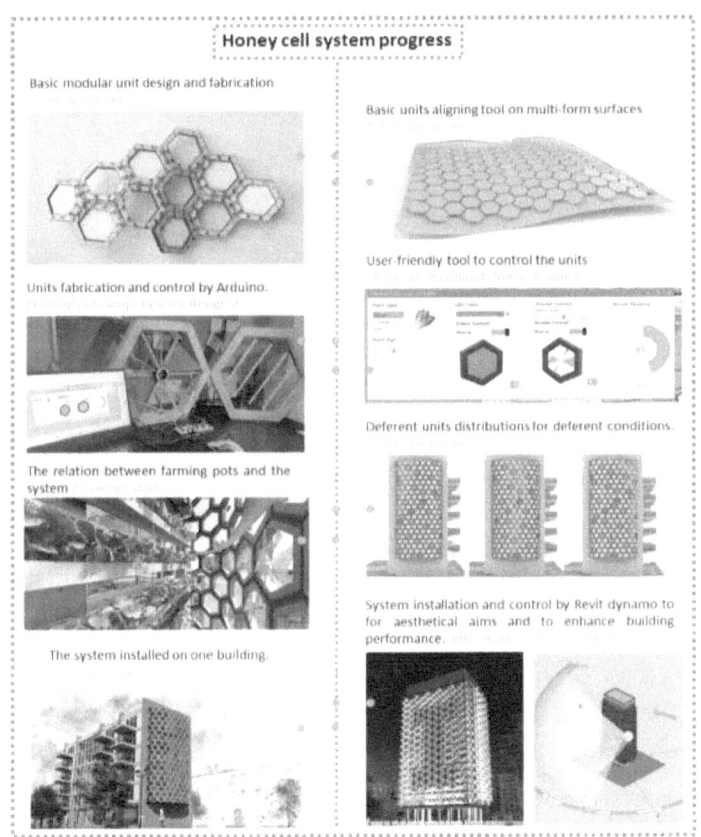

Modular hexagonal joinery system was designed, units were fabricated, controlled by Arduino, and a modular adaptive greenhouse was implemented on a building facade and a roof. Source: Ghazal et al., 2023.

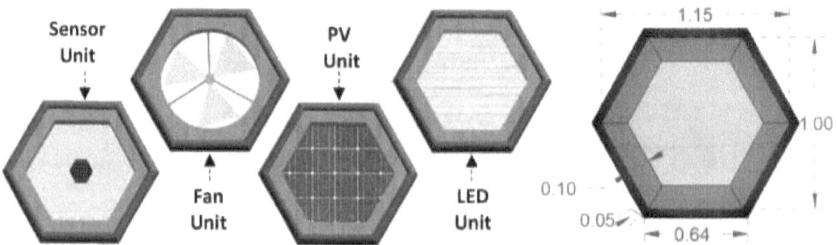

Exterior perspective for a single building from the neighbourhood. Source: Ghazal et al., 2023.

Unites types and dimensions. Source: Ghazal et al., 2023.

Medium DLI facade pattern. Source: Ghazal *et al.*, 2023.

Roof Exterior perspective for a single building from the neighbourhood.
Source: Ghazal *et al.*, 2023.

An interior perspective: how the system might look from a normal room.
Source: Ghazal *et al.*, 2023.

Integrating water and nutrient delivery systems into an existing building can be complex. Old plumbing may need to be completely overhauled to accommodate the high water demands of hydroponic or aeroponic systems. Designers must also **consider water recirculation and treatment facilities**, often necessitating the creation of new spaces within the building for these crucial components.

Climate control presents another significant challenge. Many older buildings **lack the advanced HVAC systems** required for optimal plant growth. Retrofitting often involves the installation of new climate control systems, which must be carefully integrated with the building's existing structure. This might include **creating new ductwork, installing additional insulation, or even constructing interior "buildings within buildings"** to maintain the precise environmental conditions needed for plant growth.

> **Power supply** is also a critical factor. Vertical farms, with their extensive lighting and climate control needs, often require significant electrical infrastructure upgrades.
>
> This can involve not just internal wiring changes but also negotiations with local power companies to ensure adequate supply.
>
> Some integrated farms are taking this a step further by **incorporating renewable energy systems** like rooftop solar panels, turning their high energy demand into an opportunity for sustainable power generation.

Space allocation within the building requires careful planning. Areas need to be designated not just for growing but also for seedling production, harvesting, packaging, and potentially for public interfaces like farm shops or educational spaces. This often leads to creative solutions, such as utilising basement areas for water treatment or converting former office spaces into control rooms for farm management systems.

The integration process also needs to **consider the building's existing uses and occupants.** In mixed-use scenarios, where part of the building might retain its original function while another part is converted to farming, issues of access, noise, and potential odours need to be addressed. This can lead to innovative designs that turn the farm into a feature rather than an inconvenience, such as glass-walled growing areas that serve as living art installations in office buildings.

Scientific Report

Potential and Challenges of Vertical Farming on Building Facades in Practice
Country: Australia
Publication Date: June 2024
Main focus: The study explores the integration of vertical farming (VF) on building facades, examining the benefits, challenges, and potential environmental, social, and economic impacts.

Key findings: The analysis indicates that VF on facades can reduce greenhouse gas emissions by minimising food transportation distances and improving building energy efficiency. However, high initial investment costs and potential environmental impacts from artificial lighting are significant challenges. The study notes that a well-designed VF facade can provide up to 30 tons of food per month, as demonstrated in specific case studies like the Homefarm project in Singapore.

Reference: Zhang, X., Kuru, A., Brambilla, A., & Gasparri, E. Potential and Challenges of Vertical Farming on Building Facades in Practice. *PLEA 2024 Conference Proceedings* **2024**. ResearchGate

---

Waste management is another aspect that requires careful integration. Composting facilities for plant waste might need to be incorporated, and systems for managing nutrient-rich water must be designed to comply with local regulations and avoid any negative impact on the building's existing waste management systems.

The aesthetic integration of the farm with the existing building is also important, especially in historically significant or architecturally notable structures. Designers often strive to maintain the building's external character while transforming its interior. In some cases, the vertical farm becomes a showcase feature, with viewing windows or tours that highlight the juxtaposition of historic architecture with cutting-edge agricultural technology.

> Lastly, **the regulatory landscape** must be navigated carefully.
> Zoning laws, building codes, and food safety regulations may not be designed with urban vertical farms in mind. Working closely with local authorities is often necessary to **ensure compliance** and sometimes to help shape new regulations that accommodate this innovative use of urban space.

6. Advancement in Vertical Farming

This chapter, *Advancement in Vertical Farming,* focuses on cutting-edge technologies and techniques that are pushing the boundaries of vertical farming systems.

Breeding and Genetics for Vertical Systems explores how plant breeding and genetic modifications are **tailored specifically for vertical farming environments.**

Precise Nutrient Delivery discusses strategies to **increase crop productivity** and **enhance the quality** of produce, including adjusted nutrient solutions and the use of nanomaterials.

Optimization of environmental control covers **methods for managing the growing environment**, crucial for optimising plant growth and resource efficiency in vertical farms.

Beneficial Microorganisms delve into the use of **microbial endophytes** and **Plant Growth Promoting Rhizobacteria (PGPR)**.

Artificial pollination explores **techniques for ensuring proper pollination** in enclosed vertical farming environments where natural pollinators are absent.

Breeding and Genetics

In vertical farming, plant breeding and genetics play a significant role in developing crop varieties **suited to controlled environments and limited spaces**.

Scientific Report

Controlled Environment Ecosystem: A Cutting-Edge Technology in Speed Breeding
Country: India
Publication Date: 26 June 2024
Main focus: The paper reviews the use of controlled environment ecosystems (CEEs), including vertical farming systems, in speed breeding, which is a method to accelerate plant breeding processes under carefully controlled conditions. It emphasises the role of technology, including artificial intelligence and advanced breeding techniques, in enhancing crop production and improving food security.

Key findings:
- Controlled environment systems can achieve up to 10 generations of Arabidopsis thaliana annually, significantly speeding up the breeding cycle compared to traditional methods.
- AI and machine learning are used to regulate nutrient and water efficiency, photosynthetic rates, and climate control, enhancing overall crop yield and quality.
- CEEs conserve 80% of land and 90% of water compared to traditional agriculture, while also managing nutrient runoff effectively.
- These systems improve nutrient delivery and carbon sequestration, helping to reduce greenhouse gas emissions.
- CEEs support the cultivation of a variety of crops, including biofortified crops and microgreens, addressing malnutrition and food security challenges.

Reference: Sharma, A., Hazarika, M., Heisnam, P., Pandey, H., Nampoothiri Devadas, V. A. S., Kesavan, A. K., Kumar, P., Singh, D., Vashishth, A., Jha, R., Misra, V., & Kumar, R. Controlled environment ecosystem: A cutting-edge technology in speed breeding. *ACS Omega* **2024**, 9(27), 29114-29138. DOI: 10.1021/acsomega.3c09060

Technology	Crop Breeding Application	Crop	Reference
Increased photoperiod	Increased max generations per year (from 2 to 6)	Spring barley, spring wheat	Watson et al., 2018
Increased photoperiod with accelerated vernalization	Increased max generations per year (from 2 to 5)	Winter barley, winter wheat	Cha et al., 2023
LEDs for optimal light intensity and quality	Shorten generation time with early seed harvest	Chickpea, canola	Hickey et al., 2019
Microbial inoculation	Use of *Mossene spp.* and *Rhizophagus irregularis* to defend against the root rot-causing pathogen *Aphanomyces euteiches*	Legumes	Hilou et al., 2014
Microbial inoculation	Use of *Rhizobium leguminosarum bv.* to reduce the occurrence of *Pythium* damping-off	Lentil, pea	Huang and Erickson, 2007
Microbial inoculation	Enhance growth rate by triggering the production of indole-3-acetic acid	False fleabane (*Pulicaria incisa*)	Faust and Logan, 2018
CRISPR-Cas9	Expedited production of hybrid seed lines through male sterility	Wooly tomato	Liu et al., 2021
CRISPR-Cas9	Induction of synthetic apomixis through colchicine-mediated mutagenesis	Hybrid Egyptian rice	Gaafar et al., 2017
CRISPR-Cas9	Transgenerational expression of phenotypes through synthetic apomixis induction	Hybrid rice	Liu et al., 2023
Genomic selection	Reduction of breeding program selection cycle from 19 years to 6 years	Oil palm	Shamshad and Sharma, 2018
Genomic selection	Prediction of cooking time	Common bean	Diaz et al., 2011
RNA interference	Conferral of bacterial resistance	Citrus (*Rutaceae*) family	Enrique et al., 2011
RNA interference	Conferral of fungal resistance	Rice	Jiang et al., 2009
RNA interference	Conferral of Insect resistance	Corn	Baum et al., 2007
RNA interference	Conferral of polygenic drought tolerance	Rice (Manavalan et al., 2012), cucumber (Wang et al., 2018)	Manavalan et al., 2012; Wang et al., 2018
RNA interference	Conferral of polygenic heat stress tolerance	Thale cress	Guan et al., 2013
RNA interference	Conferral of polygenic salinity tolerance	Rice	Wu et al., 2021
RNA interference	RNAi-induced silencing of MSH1 gene	Sorghum	Ketumile et al., 2022
DNA methylation	Methylation stability altered by environmental conditions induces stress resistance over generations of salinity stress conditions	Thale cress	Jiang and Doudna, 2017
DNA methylation	Methylation stability altered by environmental conditions induces stress resistance over generations of drought conditions	Rice	Zheng et al., 2017

List of utilisation examples of innovative plant breeding techniques.
Source: Williams *et al.*, 2024

At the heart of this genetic revolution is the quest for **compact growth habits**. Breeders are focusing their efforts on creating plants with **shorter internodes** and **more condensed structures**, allowing for higher planting density and optimal use of vertical space.

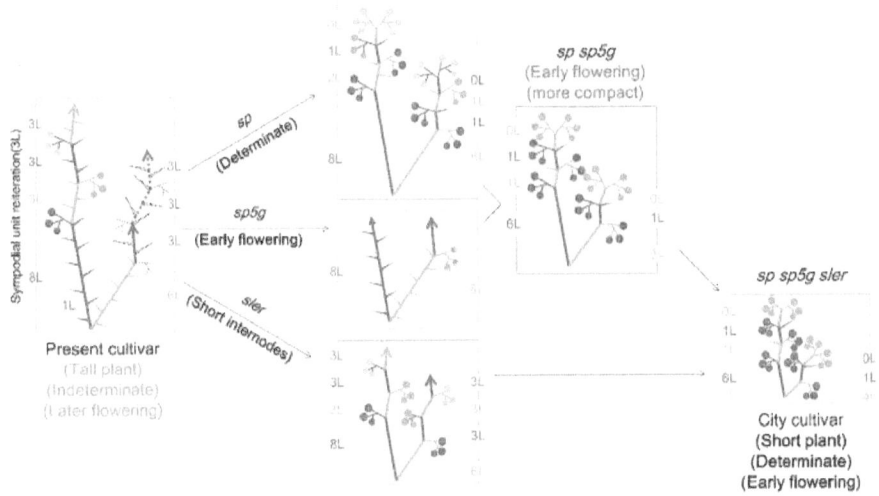

Schematic Diagram of Urban Tomato Creation via CRISPR/Cas9-Mediated Genome Editing.
Source: Fernie & Yan, 2020

Scientific Report

Targeting Key Genes to Tailor Old and New Crops for a Greener Agriculture
Country: China
Publication Date: March 2020
Main focus: The study explores the potential of CRISPR/Cas9-mediated genome editing and other genetic techniques to improve crop traits for urban agriculture, focusing on compact growth, early yield, and enhanced sustainability.

Key findings: Researchers demonstrated that genome editing could restructure vine tomato plants into compact, early-yielding varieties suitable for high-density urban farming. The study also highlighted the broader applicability of these techniques to other crops, emphasising the potential for enhancing sustainability, disease resistance, and nutritional quality.

Reference: Fernie, A. R., & Yan, J. Targeting Key Genes to Tailor Old and New Crops for a Greener Agriculture. *Molecular Plant* **2020**, 13(3), 354-356. DOI: 10.1016/j.molp.2020.02.007

Speed is another important factor driving genetic innovation in this field. By shortening the time from seed to harvest, breeders are enabling **more crop cycles per year**, increasing overall yield and profitability.

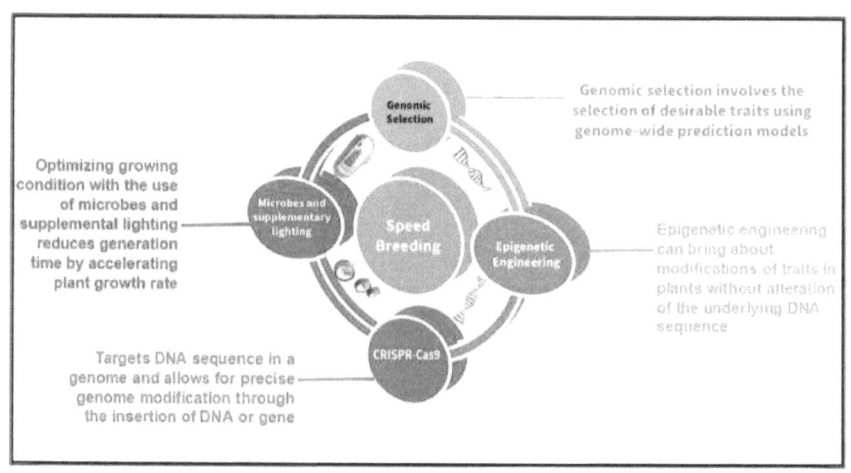

Plant breeding tools and techniques that can be utilised in speed breeding programs for rapid crop improvement. Source: Williams et al., 2024

This acceleration of natural growth cycles is pushing the boundaries of what's possible in agriculture, allowing for **year-round production**, not available historically.

Scientific Report

Technology of Plant Factory for Vegetable Crop Speed Breeding
Country: China
Publication Date: 11 July 2024
Main focus: The study explores the use of plant factories and speed breeding (SB) technologies to accelerate the breeding cycle of vegetable crops, enhancing production efficiency.

Key findings: The research highlights the advantages of using controlled environments in plant factories, such as precise control over light, temperature, and CO_2 levels, which can significantly shorten breeding cycles, potentially reducing the time required to develop new varieties by 30-50%. The integration of SB with technologies like genome editing and high-throughput genotyping presents a promising approach to rapidly develop new crop varieties with improved traits.

Reference: He R, Ju J, Liu K, Song J, Zhang S, Zhang M, Hu Y, Liu X, Li Y and Liu H (2024) Technology of plant factory for vegetable crop speed breeding. Front. Plant Sci. 15:1414860. DOI:10.3389/fpls.2024.1414860

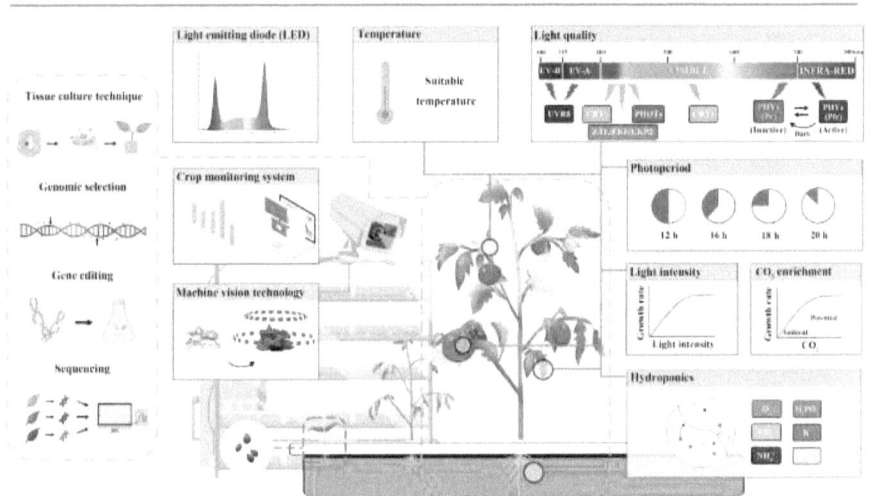

Vegetable speed breeding in a plant factory. Source: He *et al.*, 2024

Scientific Report

Tools and Techniques to Accelerate Crop Breeding
Country: United States
Publication Date: 31 May 2024
Main focus: This review focuses on modern tools and techniques used to accelerate crop breeding, such as speed breeding, CRISPR-Cas9 technology, and genomic selection, to meet the increasing demands for food security and crop improvement.

Key findings: The implementation of speed breeding can increase the number of crop generations per year from 2 to 6, significantly enhancing the breeding cycle efficiency. The use of CRISPR-Cas9 technology has been successfully demonstrated in creating male-sterile hybrid tomato lines, and genomic selection has reduced breeding cycles in crops like oil palm from 19 years to just 6 years.

Reference: Williams, K., Subramani, M., Lofton, L. W., Penney, M., Todd, A., & Ozbay, G. Tools and Techniques to Accelerate Crop Breeding. *Plants* **2024**, 13(1520).
DOI: 10.3390/plants13111520

Light efficiency is perhaps one of the most fascinating areas of genetic development for vertical farming. Plants are being bred to thrive under the artificial glow of LED lights, their photosynthetic processes fine-tuned to make the most of specific light spectra and intensities. This is not just about survival, but about flourishing in an environment that bears little resemblance to the open fields of traditional farming.

Scientific Report

How Crop Breeding Programs Can Improve Plant Factories' Business and Environmental Sustainability: Insights from a Farm Level Analysis
Country: Singapore
Publication Date: 26 December 2023
Main focus: The study analyses the potential of crop breeding programs to enhance the profitability and environmental sustainability of plant factories, using a system dynamics model to simulate various scenarios.

Key findings: The research indicates that crop breeding can significantly impact plant factory operations, with yield improvement being the most critical factor, potentially increasing profitability by up to 43 million SGD over 20 years. The study also highlights the high energy usage of plant factories, with lighting costs constituting 50% of total expenses and contributing substantially to the carbon footprint, estimated at 21 kg CO_2-eq per kg of kale produced.

Reference: Song, S., Ong, E. J. K., Lee, A. M. J., & Chew, F. T. How crop breeding programs can improve plant factories' business and environmental sustainability: Insights from a farm level analysis. *Sustainable Production and Consumption* **2024**, 44, 298-311.
DOI: 10.1016/j.spc.2023.12.020

In the soilless world of hydroponics and aeroponics, **nutrient uptake efficiency** becomes paramount. Breeders are developing varieties that can absorb and utilize nutrients more effectively in these systems, leading to reduced resource use and improved plant health. It's a delicate balance, ensuring that plants can thrive in these highly controlled environments without becoming overly dependent on conditions.

Simultaneously **biofortification** is one of the emerging topics for breeding of crops for vertical farming.

Scientific Report

Prospects of Microgreens as Budding Living Functional Food: Breeding and Biofortification through OMICS and Other Approaches for Nutritional Security
Country: India
Publication Date: 25 January 2023
Main focus: The study focuses on the potential of microgreens as nutrient-dense functional foods, exploring breeding and biofortification strategies to enhance their nutritional content and health benefits.

Key findings: The study found that microgreens are rich in essential nutrients and bioactive compounds, with some species showing up to 9.1 times higher copper content compared to mature plants. The use of OMICS approaches can further enhance the nutritional profile and yield of microgreens.

Reference: Gupta, A., Sharma, T., Singh, S. P., Bhardwaj, A., Srivastava, D., & Kumar, R. Prospects of microgreens as budding living functional food: Breeding and biofortification through OMICS and other approaches for nutritional security. Frontiers in Genetics **2023**, 14, 1053810. DOI: 10.3389/fgene.2023.1053810

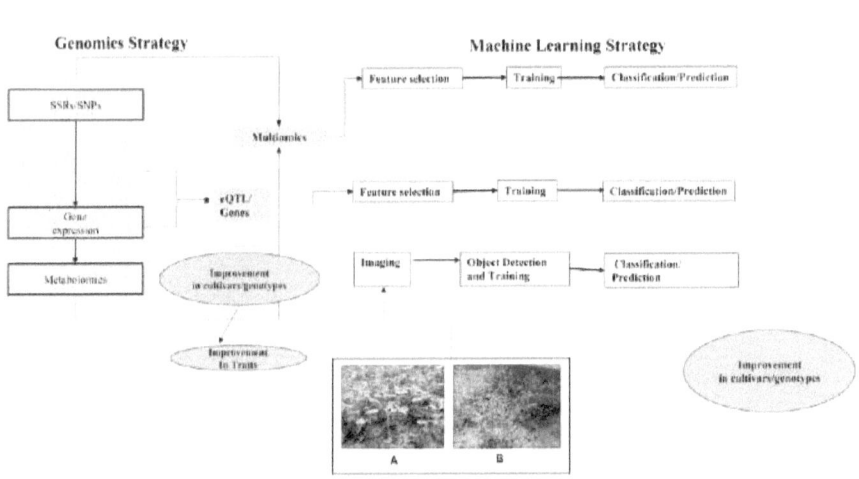

Systematic workflow of OMICS analysis and its association with machine learning for microgreens improvement and shelf-life prediction: Genomic data could be used to explore the markers in the form of SSRs, SNPs. Bulk-RNASeq and metabolomics data provides another layer of information that would enable Multi-OMICS analysis for cultivation improvement. Machine learning strategies that make the use of both "numeric" and "imaging" data would enable the development of prediction and classification models.
Source: Gupta *et al.*, 2023

Precise Nutrient Delivery

Adjustment of nutrient solutions and the **application of nanomaterials for nutrient delivery** are just a few of the ways to maximise yields and quality in vertical farming.

> **Adjusted nutrient solutions** in vertical farming are specially mixed water and nutrients designed to meet the needs of plants grown without soil. Optimised nutrient formulations to **match the specific needs of different crops** and growth stages can greatly improve growth and yield.

This includes managing pH and electrical conductivity (EC) to ensure nutrients are readily available and prevent nutrient lockout.

The optimum range of electrical conductivity and pH of common hydroponically grown crops. Source: Maluin *et al.*, 2021

Crops	pH	EC (d/Sm−1)
African Violet (*Saintpaulia ionantha* H. Wendl.)	6.0–6.8	1.4–1.8
Asparagus (*Asparagus officinalis* L.)	6.0–7.0	6.0–6.8
Banana (*Musa acuminata* Colla)	5.5–6.5	1.8–2.2
Basil (*Ocimum basilicum* L.)	5.5–6.0	1.0–1.6
Bean (*Phaseolus vulgaris* L.)	6.0	2.0 to 4.0
Broccoli (*Brassica oleracea* L. var. *italica*)	6.0 to 6.8	2.8 to 3.5
Cabbage (*Brassica oleracea* L.)	6.5 to 7.0	2.5 to 3.0
Carnation (*Dianthus caryophyllus* L.)	6.0	2.0 to 3.5
Celery (*Apium graveolens* L.)	6.5	1.8 to 2.4
Cucumber (*Cucumis sativus* L.)	5.0 to 5.5	1.7 to 2.0
Eggplant (*Solanum melongena* L.)	6.0	2.5 to 3.5
Fig (*Ficus benjamina* L.)	5.5 to 6.0	1.6 to 2.4

Crops	pH	EC (d/Sm−1)
Leek (*Allium porrum* L.)	6.5 to 7.0	1.4 to 1.8
Lettuce (*Lactuca sativa* L.)	6.0 to 7.0	1.2 to 1.8
Marrow (*Cucurbita pepo* L.)	6.0	1.8 to 2.4
Okra (*Abelmoschus esculentus* L.)	6.5	2.0 to 2.4
Pak Choi (*Brassica rapa* L.)	7.0	1.5 to 2.0
Peppers (*Capsicum annuum* L.)	5.5 to 6.0	0.8 to 1.8
Parsley (*Petroselinum crispum* (Mill.) Fuss)	6.0 to 6.5	1.8 to 2.2
Rhubarb (*Rheum* × *rhabarbarum* L.)	5.5 to 6.0	1.6 to 2.0
Rose (*Rosa abietina* Gren.)	5.5 to 6.0	1.5 to 2.5
Spinach (*Spinacia oleracea* L.)	6.0 to 7.0	1.8 to 2.3
Strawberry (*Fragaria ananassa* L.)	6.0	1.8 to 2.2
Sage (*Salvia officinalis* L.)	5.5 to 6.5	1.0 to 1.6
Tomato (*Solanum lycopersicum* L.)	6.0 to 6.5	2.0 to 4.0
Zucchini (*Cucurbita pepo* L.)	6.0	1.8 to 2.4

Dynamic nutrient solutions that change composition based on the plant's growth stage can further enhance efficiency, while real-time monitoring and adjustments using sensors and automated systems ensure that plants receive optimal nutrition continuously.

Scientific Report

Organic-Based Nutrient Solutions for Sustainable Vegetable Production in a Zero-Runoff Soilless Growing System

Country: United Arab Emirates
Publication Date: 7 February 2024
Main focus: The study evaluates the effectiveness of organic-based nutrient solutions compared to conventional inorganic fertilizers in hydroponic systems for growing lettuce, focusing on yield, nutritional content, and environmental sustainability.

Key findings: The study found that lettuce grown with organic nutrient solutions had a slightly lower yield, with an average fresh weight of 159.8 g for agro-fish liquid organic fertiliser compared to 175.1 g for inorganic fertiliser. However, the organically grown lettuce had higher total phenolic content (up to 3.36 g/100 g) and lower nitrate levels (ranging from 3.2 to 8.7 mg/kg) compared to inorganic methods.

Reference: Alneyadi, K.S.S., Almheiri, M.S.B., Tzortzakis, N., & Di Gioia, F. Organic-Based Nutrient Solutions for Sustainable Vegetable Production in a Zero-Runoff Soilless Growing System. *Journal of Agriculture and Food Research* **2024**, 15, 101035. DOI: 10.1016/j.jafr.2024.101035

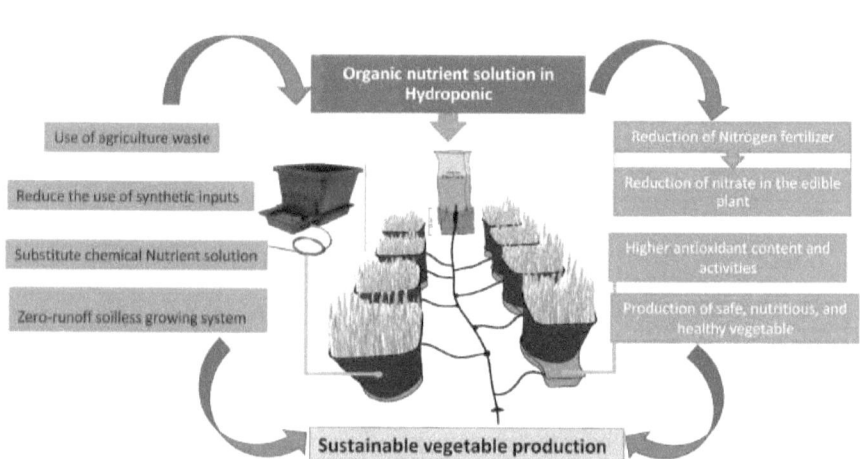

Example of soilless system with zero-runoff auto-pot technology, each unit including a 100 L tank, eight 25 L pots, main pipe and microtubes connected to a float to control the water flow in a container under each pot. Source: Alneyadi et al., 2024

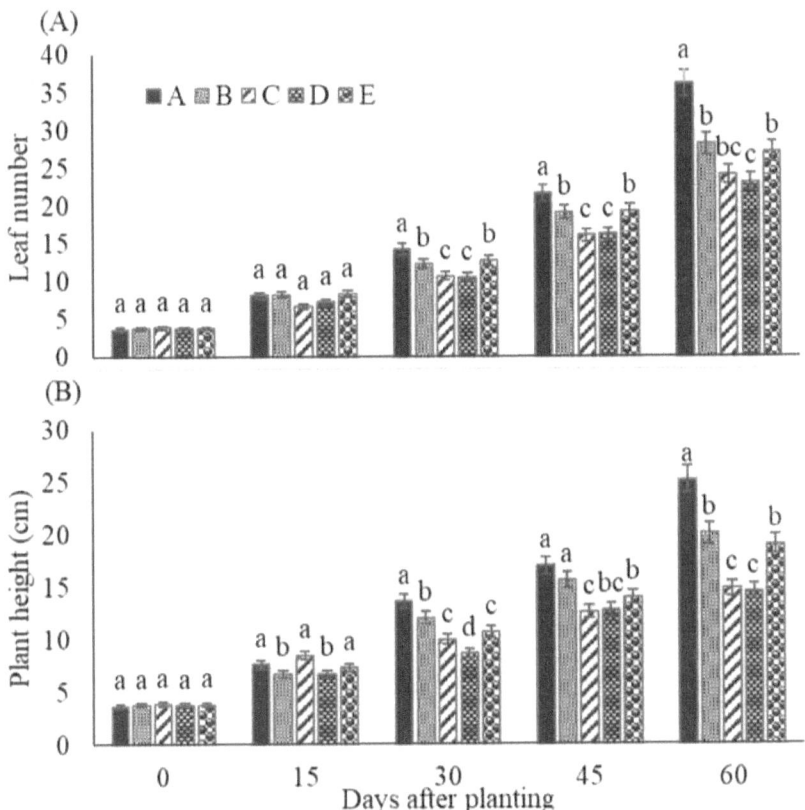

Effect of nutrient solutions (NS) in a zero-runoff auto-pot soilless system on the leaf number (A) and plant height (B) of lettuce (cultivar 'Parris Island cos'). Mean ± SE with different lowercase letters within a time interval indicate significant differences at $p ≤ 0.05$ as determined by the LSD test. a = Chemical nutrient solution (control), b = Agro-fish liquid organic fertilizer, c = Nutrihumate liquid organic fertilizer, d = Rods-fert liquid organic fertilizer, e = Bio-green liquid organic fertilizer. Source: Alneyadi et al., 2024

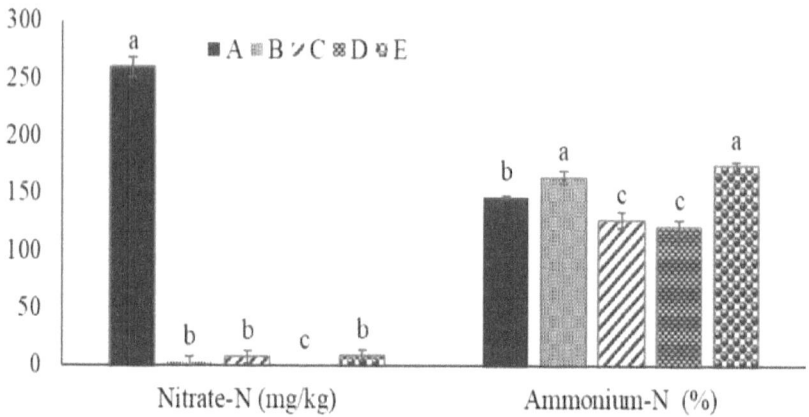

Effect of NS composition on leaf nitrate and ammonium content of lettuce (cultivar 'Parris Island cos') grown in a zero-runoff auto-pot soilless system. Mean ± SE with different lowercase letters indicate significant differences at p ≤ 0.05 as determined by the LSD test. a = Chemical nutrient solution (control), b = Agro-fish liquid organic fertilizer, c = Nutrihumate liquid organic fertilizer, d = Rods-fert liquid organic fertilizer, e = Bio-green liquid organic fertilizer. Source: Alneyadi et al., 2024

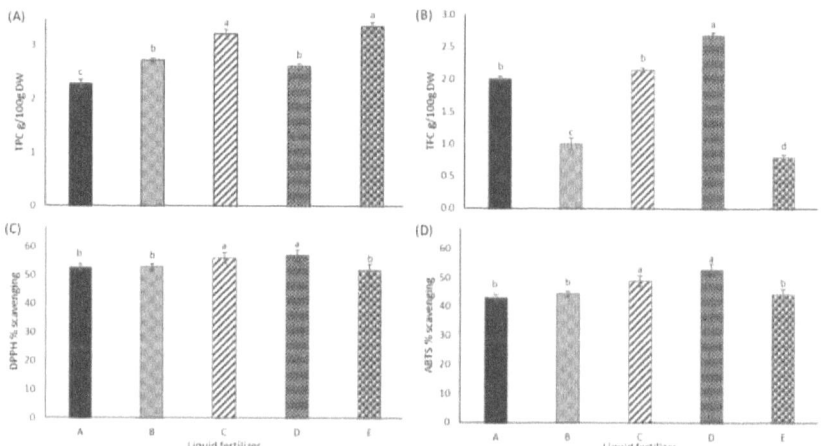

Effect of NS on total phenolics content (A), total flavonoid content (B), and antioxidant activity (C and D) of lettuce (cultivar 'Parris Island cos') leaves grown in a zero-runoff auto-pot soilless system using organic and inorganic liquid fertilizers. Mean ± SE with different lowercase letters indicate significant differences at p ≤ 0.05 as determined by the LSD test. a = Chemical nutrient solution (control), b = Agro-fish liquid organic fertilizer, c = Nutrihumate liquid organic fertilizer, d = Rods-fert liquid organic fertilizer, e = Bio-green liquid organic fertilizer. Source: Alneyadi et al., 2024

Scientific Report

Comprehensive characterization of selected phytochemicals and minerals of selected edible halophytes grown in saline indoor farming for future food production
Country: Germany
Publication Date: 3 June 2023
Main focus: This study evaluates the performance and nutritional properties of five halophyte species grown in saline indoor farming conditions, assessing their potential as sustainable food sources in response to challenges like water scarcity and climate change.

Key findings:
- The study found significant variability in nutritional content based on salinity levels. For instance, the total carotenoid content ranged from 1581.4 ± 180.4 ng mg-1 DW in scurvy grass to 188.3 ± 48 ng mg-1 DW in glasswort with no salt.
- Glasswort exhibited a 40-fold increase in β-carotene under 200 mM salt compared to other halophytes.
- Chloride content in palm kale increased 20-fold at 200 mM salt, and nitrate levels decreased by 1.5-2 times from 200 mM to 1200 mM salt in garden orache and scurvy grass.

Reference: Fitzner M., Schreiner M., Baldermann S., Comprehensive characterization of selected phytochemicals and minerals of selected edible halophytes grown in saline indoor farming for future food production, *Journal of Food Composition and Analysis* **2023**, Volume 122, 105435. DOI: 10.1016/j.jfca.2023.105435.

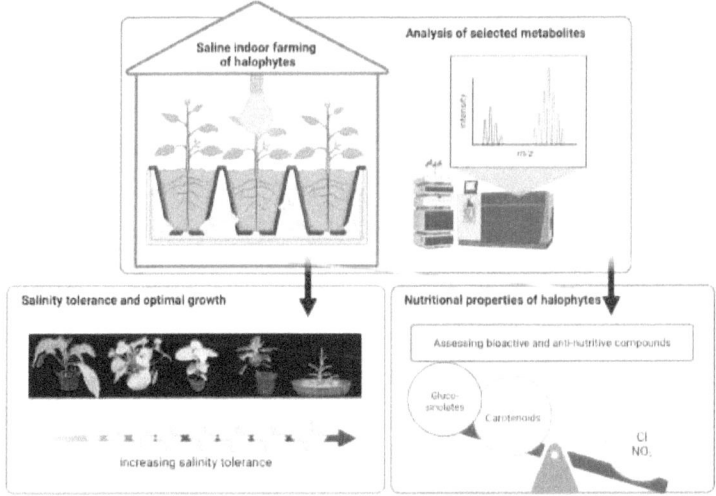

Schematic description of characterization of selected phytochemicals and minerals of edible halophytes, including palm kale, scurvy grass, quinoa, garden orache, and glasswort, grown in saline indoor farming. Source: Fitzner *et al.*, 2023

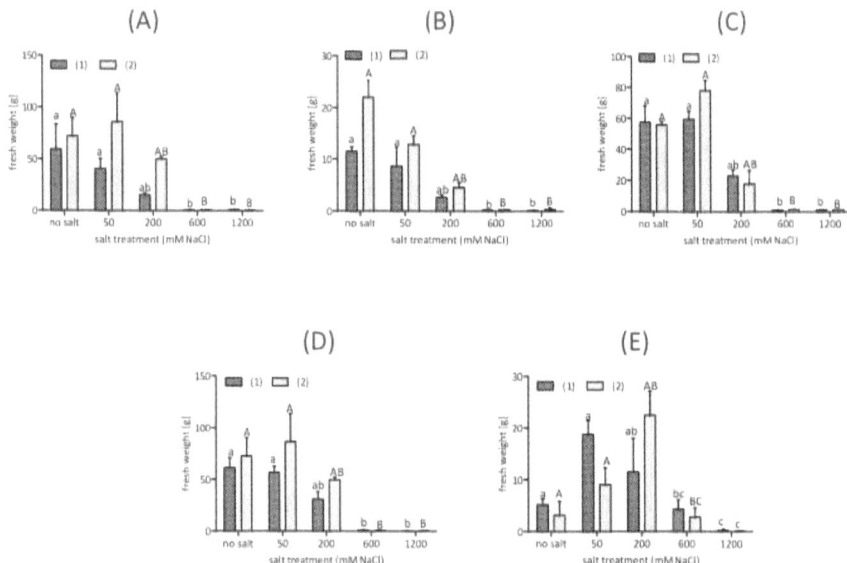

Effect of salt treatment on the fresh weight of 6 or 9-week-old plants depending on the plant species. (A) Palm kale; (B) scurvy grass; (C) quinoa; (D) garden orache; (E) glasswort. The average fresh weight and standard deviation were measured in two experimental setups (1) and (2). Lowercase letters indicate significant differences between treatments in setup 1, sorted from highest to lowest; uppercase letters indicate significant differences between treatments in setup 2, also sorted from highest to lowest, ($p \leq 0.05$).
Source: Fitzner et al., 2023

The **application of nanomaterials** for nutrient delivery offers additional benefits by enhancing nutrient efficiency and improving plant health.

Nano-fertilizers, with their **high surface area and reactivity**, increase nutrient availability and uptake, providing a controlled release of nutrients that ensures a consistent supply. These materials can also **enhance disease resistance and stress tolerance**

Environmentally, nanomaterials help **reduce chemical use** and nutrient leaching. Integrating these advanced techniques into vertical farming systems results in higher yields, better quality produce, and more sustainable farming operations.

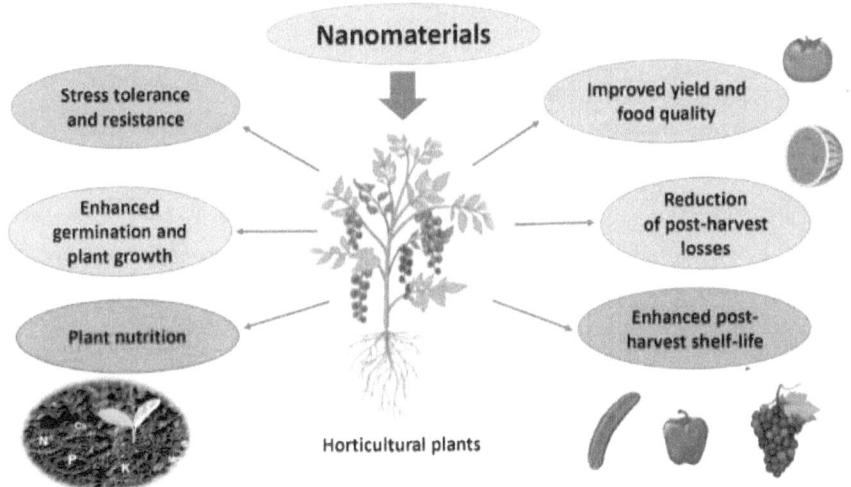

Importance of nanomaterials in horticultural crop production. Source: Lastochkina et al., 2022

Scientific Report

Some Emerging Opportunities of Nanotechnology Development for Soilless and Microgreen Farming

Country: Malaysia
Publication Date: June 15, 2021
Main Focus: This review explores the potential applications of nanotechnology in enhancing soilless and microgreen farming, aiming to provide more sustainable and productive alternatives to conventional farming methods.

Key Findings:

1. Nutrient Nanoparticles can minimise nutrient losses and improve nutrient uptake and bioavailability in crops.
2. Nano-sensing and advanced sensors can provide real-time detection of pH, temperature, and nutrient levels, allowing for precise environmental control.
3. Incorporating nanoparticles can improve the quality of substrates used In soilless farming.
4. Nanomaterials can enhance plant resistance to environmental stress and diseases.
5. Nanoparticles can be used to extend the shelf life of microgreens through innovative packaging solutions.

Reference: Maluin, F.N.; Hussein, M.Z.; Nik Ibrahim, N.N.L.; Wayayok, A.; Hashim, N. Some Emerging Opportunities of Nanotechnology Development for Soilless and Microgreen Farming. *Agronomy* **2021**, *11*, 1213. DOI: 10.3390/agronomy11061213

Some of the most recent R&D on nanotechnology approaches to improving hydroponic nutrient solutions. Source: Maulin *et al.*, 2021

The Incorporation of Nanoparticles into Nutrient Solutions	Type of Crops	Method of Soilless Cultivation	Finding
Fe_2O_3 nanoparticles (30–40 nm) at concentrations of 100, 150, and 200 mg are mixed with Hoagland nutrient solution	Spinach (*Spinacia oleracea* L.)	Hydroponic	According to the findings, adding nano Fe_2O_3 to spinach boost its growth rate in a dose- and time-dependent manner. After 45 days, the stems and roots of spinach grown in various Fe_2O_3 concentrations at 100, 150, and 200 mg, are approximately 1.45, 1.91, respectively, and 2.27 and 1.25, 1.38, and 1.75, respectively, times longer than the control spinach.
ZnO nanoparticles (25 nm) at concentrations of 0.2, 1, 5 and 25 µg are mixed with Johnson nutrient solution	Tobacco (*Nicotiana tabacum* L.)	Hydroponic	When compared to the control, Nano-ZnO increased biomass indices such as root and shoot main and lateral lengths, as well as root and shoot weight. Low or middle levels of ZnO nanoparticles increased amino acids, phenolic compounds, proline, reducing sugars, and flavonoids whereas 25 µM ZnO nanoparticles did not increase proline or flavonoids. Nano-ZnO application increased the activity of superoxide dismutase, peroxidase, glutathione peroxidase, and polyphenol oxidase more than bulk-ZnO application.
Se nanoparticles (8–15 nm) at concentrations of 1, 4, 8 and 12 µM are mixed with a nutrient solution mixture of N (116 mg L−1), P (21 mg L−1), K (82 mg L−1), Ca (125 mg L−1), Mg (21 mg L−1), S (28 mg L−1), Fe (6.8	tomato (*Solanum lycopersicum* L.)	Hydroponic	The study discovered that both bulk Se (at concentrations of 2.5, 5, and 8 µM) and Se nanoparticles (at concentrations of 4, 8, and 12 µM) had positive effects on tomato growth parameters by increasing the fresh and dry weight and diameter of the shoots, as well as the fresh and dry weight and volume of the roots. In terms of chlorophyll content of tomato leaves grown under low-temperature stress (10 °C for 24 h), Se

The Incorporation of Nanoparticles into Nutrient Solutions	Type of Crops	Method of Soilless Cultivation	Finding
mg L−1), Mn (1.97 mg L−1), Zn (0.25 mg L−1), B (0.70 mg L−1), Cu (0.07 mg L−1), and Mo (0.05 mg L−1)			nanoparticles (27.5%) outperformed bulk Se (19.2%).
SiO2 nanoparticles (20–40 nm) at a concentration of 1% w/v is mixed with Hoagland nutrient solution	Maize (*Zea mays* L.)	Hydroponic	Hydroponically grown maize absorbed SiO2 nanoparticles at a rate of 18.2%, resulting in a 95.5% increase in germination, a 6.5 % increase in dry weight, and better nutrient alleviation in seeds exposed to SiO2 nanoparticles than in seeds exposed to bulk silicon of SiO2, Na2SiO3 and H4SiO4 and control.
Zein nanoparticles (135 nm) at concentrations of 0.88 and 1.75 mg/mL are mixed with Hoagland nutrient solution	Sugar cane (*Saccharum officinarum* L.)	Hydroponic	After 12 h of exposure to zein nanoparticles, the concentration of nanoparticles adhering to sugar cane roots varied with dosage, with 110.2 μg NPs/mg dry weight of root in a low dose nanoparticle suspension (0.88 mg/mL) and 342.5 μg NPs/mg dry weight of root in a high dose nanoparticle suspension (1.75 mg/mL). The translocated nanoparticles were then observed in leaves with 4.8 μg NPs/mg dry weight of leaves in a low dose nanoparticle suspension (0.88 mg/mL) and 12.9 μg NPs/mg dry weight of leaves in a high dose nanoparticle suspension (1.75 mg/mL).
Hoagland nutrient solution was used in the early phase, and after the third leaf had fully expanded, hydroxyapatite nanoparticles (94–163 nm) at concentrations of 2, 20, 200, 500, 1000, and 2000 mg L−1 were mixed with 1% w/v carboxymethylcellulose	Tomato (*Solanum lycopersicum* L.)	Hydroponic	There were no phytotoxic effects on tomato plants grown in hydroponics with hydroxyapatite nanoparticles and increasing the concentration of the nano-mixture induces root elongation. For 200 and 500 mg L−1, the increase in root length was +64% and +97%, respectively, when compared to the control.

The Incorporation of Nanoparticles into Nutrient Solutions	Type of Crops	Method of Soilless Cultivation	Finding
Fe3O4 nanoparticles or TiO2 nanoparticles (10–30 nm) at concentrations of 50 and 500 mg/L are mixed with nutrient solution mixture of N (11.0 mM), P (1.2 mM), Ca (4.0 mM), K (7.0 mM), S (2.41 μM), Fe (17.8 μM), Zn (5.0 μM), Mn (10.0 μM) and Cu (2.7 μM)	Tomato (*Solanum lycopersicum* L.)	Hydroponic	When compared to the control and seedlings exposed to Fe3O4 nanoparticles, seedlings grown with high concentrations of TiO2 nanoparticles displayed an irregular proliferation of root hairs one week after the start of the nanoparticle treatment. Tomato seedlings grown under different conditions had similar shoot morphology, and plants treated with nanoparticles showed no signs of toxicity.
Cu-Fe2O4 nanoparticles at concentrations of 0.0, 0.04, 0.2, 1, and 5ppm are mixed with Hoagland nutrient solution	Cucumber (*Cucumis sativus* L.)	Hydroponic	After being exposed to Cu-Fe2O4 nanoparticles, cucumber plants' fresh weight and protein content increased. The activities of superoxide dismutase and peroxidase were also substantially higher in cucumber shoots and roots. The use of Cu-Fe2O4 nanoparticles improved the absorption of Fe and Cu by cucumber tissues significantly.
Chitosan nanoparticles (149 nm) or chitosan-indole-3-acetic acid nanoparticles (183 nm) at various ratio are mixed with La Molina nutrient solution	Lettuce (*Latuca sativa* L.)	Hydroponic	Hydroponically grown lettuce treated with chitosan nanoparticles and chitosan-indole-3-acetic acid nanoparticles exhibits significant increases of 42.6% and 30.9%, respectively, compared to the control. In terms of the effect on leaf size, chitosan nanoparticles outperformed other treatments with the largest leaves.

Some of the recent patents on nanotechnology approaches in aerated soilless farming. Source: Maulin et al., 2021

Patent No./Year/Title	Method of Soilless Cultivation	Invention
N102701844B/2012/Rich-selenium-germanium trace element nanometer nutrition fertilizer for vegetable and fruit soilless culture	Hydroponic	The invention describes the preparation and manufacture of nutritional fertilizer rich in selenium and germanium trace elements for vegetable and fruit cultivation in courtyards or balconies using soilless cultivation.
CN206354136U/2017/A kind of indoor micro-nano bubble hydroponic device	Hydroponic	The current utility model's cultivation cabinet is a semi-hermetic layer stereo system, with the bottom opening passage effectively carrying out indoor and cultivation cabinet air exchange with reference to the ventilation ventilating fan. Aeration will be used by the micro-nano bubble generator to generate the other micro/nano level water vapor bubbles. The amount of dissolved oxygen increases the nutrient solution essentially.
JP2015097515A/2013/Hydroponic raising seedling method, and hydroponic culture method	Hydroponic	The invention is to provide a hydroponic seedling system capable of raising a strong seedling and shortening the seedling raising period by adding a hydroponic solution containing micro-nano bubbles during the plant seedling period.
KR20130086099A/2012/The method manufacture silver nano antimicrobial & lacquer tree a composite in uses functionality crop	Hydroponic	The current innovation is a method of growing functional crops using a silver nano antibacterial agent and a lacquer composition through hydroponic cultivation.
CN105417674A/2015/Preparation method and application of micro-nano sparkling water	Hydroponic	The invention reveals a method for preparing micro-nano sparkling water, which benefits the field of scientific and technological agriculture in areas such as soilless production, fruit and vegetable washing, biological repair, dirty water processing, and so on.
WO2017101691A1/2015/The method for cultivation of plants using metal nanoparticles and the nutrient medium for its implementation	Hydroponic	Seed germination and subsequent plant cultivation on an aseptic agar nutrient medium containing a variety of organic and inorganic components important for plant growth, such as iron, zinc, and copper in the form of electro-neutral metal nanoparticles. Chitosan can also be added to the nutrient medium. This process improves seed germination as well as plant physiological and morphological indices such as root length and root behavior, chlorophyll content in leaves, sprout length, and green mass yield.

Patent No./Year/Title	Method of Soilless Cultivation	Invention
KR20060055895A/2004/Silver nano-containing bean sprouts manufacturing equipment	Hydroponic	The present invention relates to the production of bean sprouts for cultivation with silver-containing water when the bean sprouts are cultivated.
CN203482710U/2013/Oxygenation and disinfection device for soilless cultivation nutrient solution	Hydroponic	A filter, an oxygen generator, an ozone generator, a rapid micro-nano bubble generator, and an ultraviolet disinfector are all part of the soil-free nutrient solution oxygenation and disinfection system.
AU2015370052B2/2014/Nano particulate delivery system	Hydroponic and aeroponic	The invention describes a system for delivering nano lipids, more specifically a nano concentrate, a nano lipid stable emulsion, a method for preparing nano lipid concentrates, and a system for delivering lipids for use as a carrier in manufacturing, medical, animal, horticultural, and agricultural chemistry.
AU2016202162B2/2012/Plant nutrient coated nanoparticles and methods for their preparation and us	Hydroponic and aeroponic	The invention describes a nanofertilizer with at least one plant nutrient coated on a metal nanoparticle that is made by combining a metal salt and a plant nutrient in an aqueous medium and then adding a reducing agent to the solution to form a coated metal nanoparticle.
TW201902343A/2017/Fish and vegetable symbiosis system including a support, at least one planting unit, a filtering unit, and a breeding unit	Aquaponic	The invention discloses a fish and vegetable symbiosis system comprising a support, at least one planting unit, one filtering unit, and one breeding unit. For water quality filling, the fish and vegetable symbiosis device is outfitted with an artificial closed form of composite filter material-activated carbon nano silver photocatalyst.
CN104719233A/2015/Nano-catalysis aquaponics method	Aquaponic	The invention includes nano-catalyst aquaponics preparation steps involving the use of purple grit dust, tourmaline, nano-titanium, nano-magnesia, medical stone, and zeolite.

Optimization of Environmental Control

At the heart of this precision agriculture lies **environmental control**.

Advanced LED systems surround crops in light spectrums optimised for their growth stage, changing throughout the day to **mimic natural cycles** or **push for accelerated development**. Temperature management goes beyond simply maintaining warmth; it often involves creating **day-night differentials** that signal plants to grow in specific ways, much like they would respond to seasonal changes in nature.

Scientific Report

Thermal Management and Energy Efficiency Analysis of Planar-Array LED Water-Cooling Luminaires in Vertical Farming Systems for Saffron
Country: China
Publication Date: 30 September 2023
Main focus: This study evaluates the thermal management and energy efficiency of planar-array LED water-cooling luminaires specifically designed for saffron cultivation in vertical farming systems.

Key findings:
- The water-cooling system effectively reduced the maximum deviation in substrate temperature to less than 2.4°C.
- The photosynthetic photon flux efficacy (PPFE) for blue LEDs at 450 nm was 2.74 µmol/J, for red LEDs at 660 nm was 1.65 µmol/J, and for white LEDs at 3000 K was 2.75 µmol/J.
- White LED luminaires demonstrated higher PPFE and lower power consumption, making them more suitable for saffron cultivation compared to blue-red luminaires.

Reference: Gao D, Ji X, Pei W, Zhang X, Li F, Han Q, Zhang S. Thermal management and energy efficiency analysis of planar-array LED water-cooling luminaires in vertical farming systems for saffron. *Case Studies in Thermal Engineering.* **2023**, 51:103535. DOI: 10.1016/j.csite.2023.103535

(a) (b)

The actual lighting effect of LED luminaire with WCL-PS. (a) White luminaire under the blooming stage of saffron. (b) Blue-red luminaire with the B/R ratio of 20%: 80% under the flower bud differentiation stage. Source: Gao *et al.*, 2023

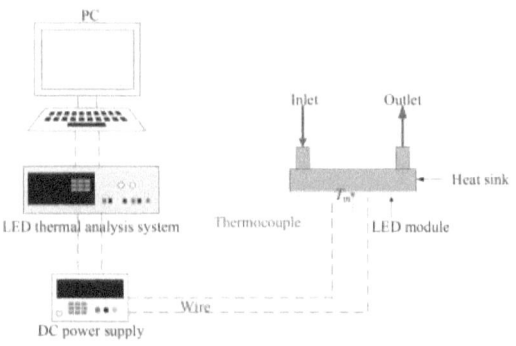

Thermal performance test platform. (a) Schematic diagram. (b) Measurement point position T_m of an LED module. (c) Measurement point positions T_{m1}–T_{m8} of WCL-PS.
Source: Gao et al., 2023

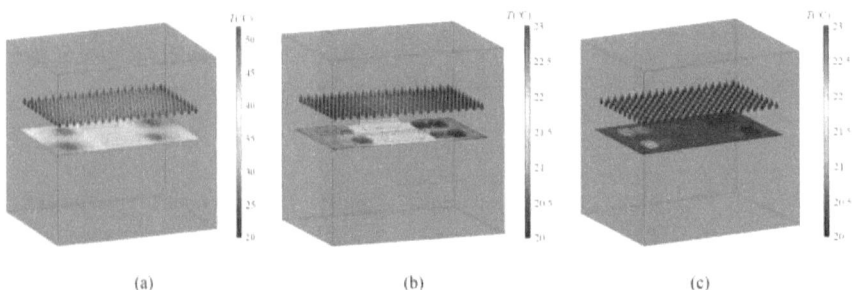

Surface temperature distribution of B-R-LED luminaires with three types of structures from numerical simulation results, within the vertical space range of ±450 mm of luminaires in the culture shelf system. (a) LED natural-cooling luminaire with power supplies and dimming control systems (NCL + PS). (b) LED water-cooling luminaire with power supplies and dimming control systems (WCL + PS). (c) LED water-cooling luminaire without power supplies and dimming control systems (WCL-PS). Source: Gao et al., 2023

Scientific Report

Optimising Growth Conditions in Vertical Farming: Enhancing Lettuce and Basil Cultivation through the Application of the Taguchi Method

Country: Iran
Publication Date: 10 May 2023
Main focus: The study focuses on optimising environmental conditions for lettuce and basil cultivation in vertical farming systems using the Taguchi method, a statistical approach for designing experiments.

Key findings:

- Taguchi Method Efficiency: The use of the Taguchi method significantly reduced the number of experiments needed, identifying the optimal conditions for plant growth with high accuracy and minimal error.
- Key factors such as electrical conductivity, LED lighting types, and CO_2 concentration were optimised. The study found that EC of 0.9 dS/m and 1.2 dS/m were ideal for lettuce and basil, respectively.
- The optimal conditions improved various growth parameters including leaf number, leaf area, biomass, and nutrient content in both lettuce and basil.
- The study highlighted the significant interrelationships between factors like light, temperature, and humidity, demonstrating their combined impact on plant growth.
- The findings provide valuable insights for enhancing crop production and sustainability in vertical farming, particularly for leafy greens like lettuce and basil.

Reference: Farhangi, H., Mozafari, V., Roosta, H.R. et al. Optimizing growth conditions in vertical farming: enhancing lettuce and basil cultivation through the application of the Taguchi method. *Sci Rep* **2023** 13, 6717. DOI: 10.1038/s41598-023-33855-z

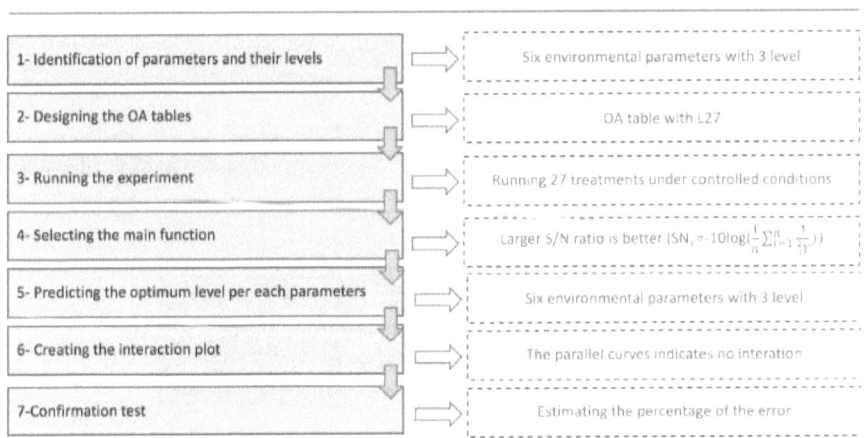

The workflow to analyse the data using Taguchi approach (S: Signal N: Noise ratio).
Source: Farhangi et al., 2023

Scientific Report

Several short-day species can flower under blue-extended long days, but this response is not universal
Country: Netherlands
Publication Date: 15 November 2023
Main focus: This study investigates whether several short-day plant species can flower under blue-extended long days, using dynamic LED lighting to enhance growth and flowering in vertical farms.

Key findings:
- The study found that artemisia, chrysanthemum, cosmos, poinsettia, and wild tomato flowered under blue-extended long days, while kalanchoe, perilla, and stevia did not.
- Plants under blue-extended long days received 15% higher daily light integral (DLI), resulting in a 4-36% increase in total dry weight compared to short days.
- Additionally, internode length increased by 6-48% under blue-extended long days compared to short days.

Reference: Sharath Kumar M., Luo J., Xi Y., van Ieperen W., Marcelis L.F.M., Heuvelink E., Several short-day species can flower under blue-extended long days, but this response is not universal, *Scientia Horticulturae* **2024**, 325, 112657, DOI: 10.1016/j.scienta.2023.112657

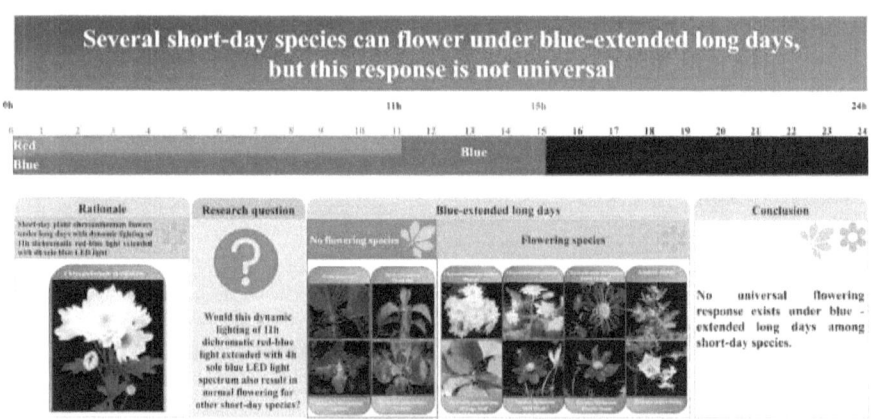

Several short-day species can bloom in blue-extended long days, but not all.
Source: Sarath Kumar *et al.*, 2024

Humidity control is a delicate balance, crucial for preventing diseases while promoting robust growth. Many vertical farms take this a step further by enriching the air with CO2, improving photosynthesis and boosting growth rates.

The precision these systems offer is incredible. In a well-designed vertical farm, operators can **create microclimates within the larger growing space**, tailoring conditions for different crops or even different stages of growth within the same crop.

Scientific Report

Analysis of Cross-Influence of Microclimate, Lighting, and Soil Parameters in the Vertical Farm

Publication Date: 19 August 2023

Main focus: This study investigates the mutual influences of microclimate parameters, lighting, and soil conditions on potato growth within a controlled vertical farm environment.

Key findings:
- The correlation coefficient between soil moisture and carbon dioxide concentration was found to be 0.42, with the best match occurring after a 55.8-hour shift, indicating a delayed response in CO2 absorption following irrigation.
- Increased peat content in the substrate shifts the dominant influence on soil moisture from air humidity to air temperature, enhancing the evaporation process and affecting photosynthesis.
- The highest mean absolute percentage error (MAPE) for predictive models of soil moisture content was 1.78%, with the best-performing model having an R^2 value of 0.821.

Reference: Kamenchuk, V.; Rumiantsev, B.; Dzhatdoeva, S.; Sadykhov, E.; Kochkarov, A. Analysis of Cross-Influence of Microclimate, Lighting, and Soil Parameters in the Vertical Farm. Agronomy **2023**, 13, 2174. DOI: 10.3390/agronomy13082174

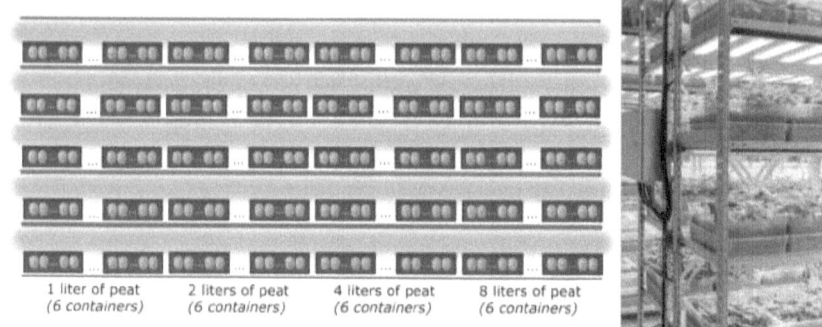

Structure of a part of the vertical farm. Source: Kamenchuk et al., 2023

The lighting conditions used for potato cultivation.
Source: Kamenchuk et al., 2023

Crop	Variety	Step	Duration, Days	Blue, %	White, %	Red, %	Turn-On Time	Turn-Off Time	PAR, µmol/(s·m2)
Potato	Innovator	1	Sep 20–Sep 29	0	0	0	0:00	0:00	0
		2	Sep 30–Oct 24	30	70	0	8:00	22:00	220
		3	Oct 25–Nov 8	0	70	50	8:00	22:00	232
		4	Nov 9–Nov 28	30	70	50	8:00	22:00	299
		5	Nov 29–Dec 9	30	70	0	8:00	22:00	220

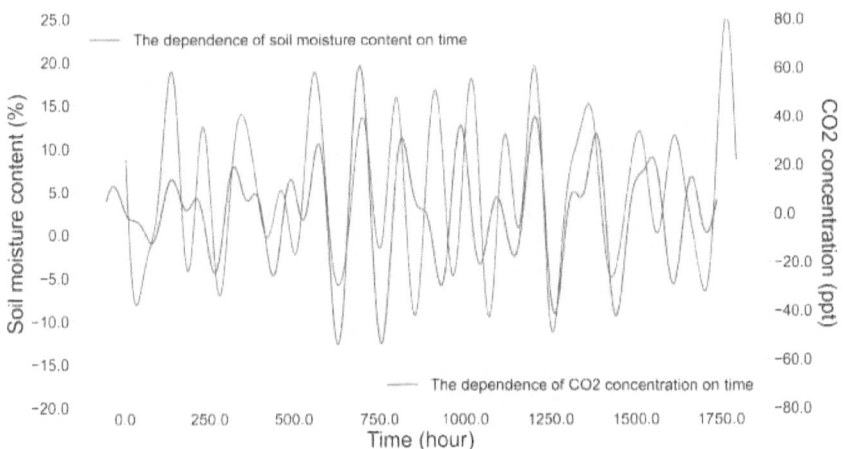

Filtered dependencies of the soil moisture content and shifted CO2 concentration on time.
Source: Kamenchuk et al., 2023

Some researchers are exploring the **use of plant-based sensors** - devices that can directly measure plant stress responses and feed that information back into the climate control system.

Some are exploring the integration of unconventional environmental factors, such as **manipulating air ionisation** and **sound vibrations**, to achieve optimal growing conditions.

Scientific Report

Effects of Air Anions on Growth and Economic Feasibility of Lettuce: A Plant Factory Experiment Approach

Country: Republic of Korea
Publication Date: 21 November 2022
Main focus: This study investigates the effects of air anions on the growth of lettuce and evaluates the economic feasibility of this technique within a plant factory setting.

Key findings:

- The application of negatively ionised air improved lettuce growth significantly.
- For red leaf lettuce, the leaf area increased by 1.3 times, shoot fresh weight increased by 1.5 times, and dry weight increased by 1.2 times.
- For Lollo bionda lettuce, the leaf area and fresh weight increased by 1.5 times each.
- The economic analysis revealed that the annual net profit per 1500 m² was approximately USD 60,000 for red leaf lettuce and USD 70,000 for Lollo bionda lettuce.

Reference: Lee, S.; Song, M.-J.; Oh, M.-M. Effects of Air Anions on Growth and Economic Feasibility of Lettuce: A Plant Factory Experiment Approach. *Sustainability* **2022**, *14*, 15468. DOI: 10.3390/su142215468

Application of air anions to lettuce cultivation in a commercial plant factory (**A**).
Source: Lee *et al.*, 2022

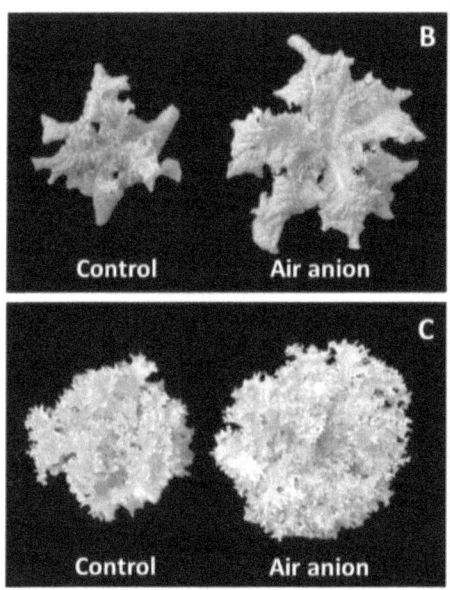

Red leaf (**B**) and Lollo bionda lettuce (**C**) plants grown under the application of air anions for 28 days.
Source: Lee *et al.*, 2022

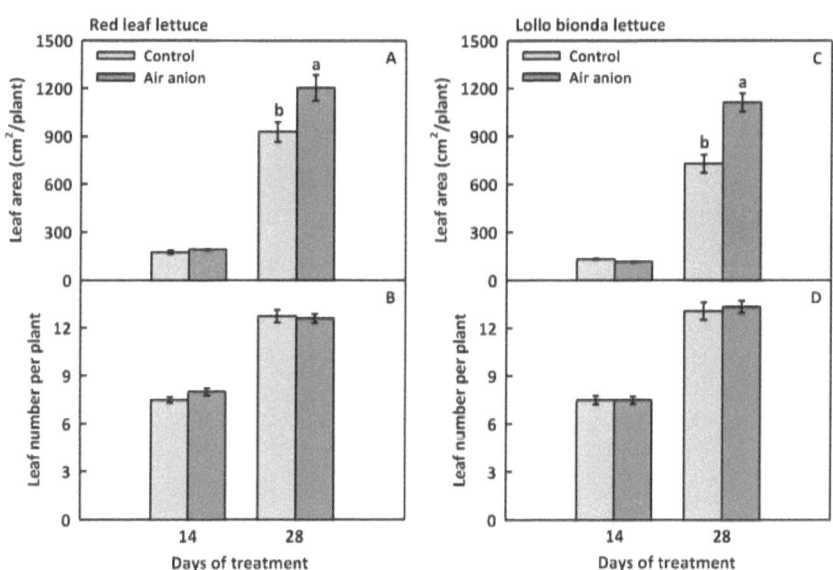

Leaf area (**A,C**) and number (**B,D**) of two types of lettuce grown under the application of air anions (5 × 10⁵ ions·cm−3) for 14 and 28 days. Different lowercase letters indicate significant differences at $p < 0.001$ (n = 15). Source: Lee *et al.*, 2022

Expenses, yield, and revenue after applying air anions to the cultivation of two cultivars of lettuce in a plant factory. Source: Lee *et al.*, 2022

Type of Lettuce	Treatment	
	Control	Air Anion
Red leaf		
Expenses (USD)	-	26.3
Yield (kg)	14.5	21.3
Revenue (USD)	-	108.8
Lollo bionda		
Expenses (USD)	-	26.3
Yield (kg)	14.2	21.6
Revenue (USD)	-	125.8

Cultivation period: 4 weeks (1 harvest); cultivation area: 15 m2; a set of air anion generators installed per unit area (15 m2); as expenses we considered those associated with the generators and consumed electricity; lettuce price: USD 1.6 and 1.7 per 100 g for red leaf and Lollo bionda lettuce, respectively; exchange rate applied as of 5 July 2022 (KRW 1300/USD 1).

Beneficial Microorganisms

Beneficial microorganisms are almost invisible but they play an important role in vertical farming.

These microorganisms **enhance plant growth, nutrient uptake, and resilience to environmental stressors**.

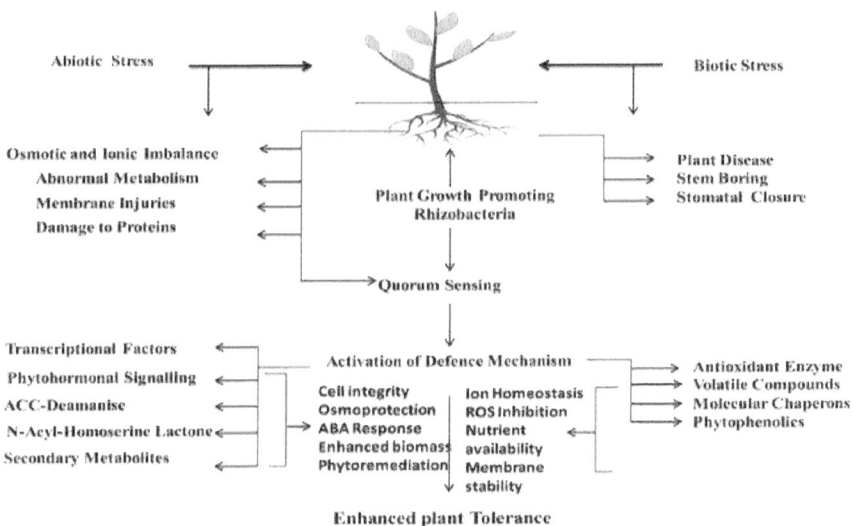

Mechanism and protection against stresses by PGPR. Source: Khan et al., 2020

Beneficial microorganisms, including bacteria and fungi, help to create a more sustainable and efficient farming system by **reducing the need for chemical fertilizers and pesticides**.

Scientific Report

Novel Approaches for Sustainable Horticultural Crop Production: Advances and Prospects
Country: Iran
Publication Date: 5 October 2022
Main focus: The study reviews innovative techniques and tools to enhance the sustainable production of horticultural crops, focusing on microbial endophytes, nanomaterials, strigolactones, CRISPR, and controlled environment horticulture.

Key findings: The research highlights the use of microbial endophytes to improve crop yield and stress tolerance, with evidence showing a 20-50% increase in crop yield under various stress conditions. It also discusses the potential of nanomaterials and controlled environment horticulture (CEH) to optimise resource use and enhance crop quality.

Reference: Lastochkina, O.; Aliniaeifard, S.; SeifiKalhor, M.; Bosacchi, M.; Maslennikova, D.; Lubyanova, A. Novel Approaches for Sustainable Horticultural Crop Production: Advances and Prospects. *Horticulturae* **2022**, *8*, 910. DOI: 10.3390/horticulturae8100910

Microbial endophytes provide beneficial effects on plants by interacting with various plant hormones and enzymes. These include abscisic acid (ABA), cytokinin (CK), gibberellins (GA), ethylene (Et), and jasmonic acid (JA), which regulate growth and stress responses. They also help with nutrient uptake, like nitrogen (N) and phosphorus (P), and enhance plants' defense mechanisms through induced systemic resistance (ISR) and systemic acquired resistance (SAR). Additionally, they assist in combating oxidative stress with enzymes like ascorbate peroxidase (APX), catalase (CAT), and superoxide dismutase (SOD), and produce beneficial compounds like lipopeptides (LPs) and volatile organic compounds (VOCs). Source: Lastochkina *et al.*, 2022

Bacterial inoculation is applicable for substrate-grown plants in vertical farming. It can enhance nutrient availability, promote plant growth, and improve disease resistance by introducing beneficial bacteria to the root zone, which can be particularly effective in controlled environments like vertical farms.

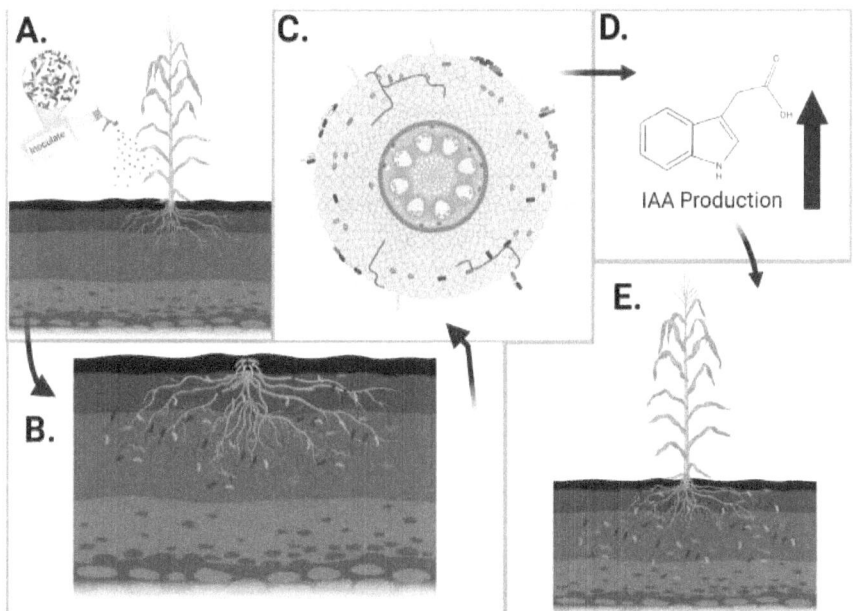

Bacterial inoculation of nutrient-deficient plants. (A) Soil growing a corn plant exhibiting nitrogen deficiency symptoms is spray-inoculated with PGP bacterial cocktail. (B) Bacterial colonies distribute throughout soil, then attach to and penetrate root tissue (C). This results in an increase in indole-3-acetic acid production within the plant (D), which supports plant growth mechanisms, targets the deficiency, and alleviates the symptoms (E).
Source: Williams et al., 2024

Plant Growth Promoting Rhizobacteria (PGPR) are applicable in soilless agriculture through their integration into hydroponic and aeroponic nutrient solutions.

PGPR **colonises plant root surfaces and internal tissues**, providing benefits such as improved nutrient solubilization and uptake, phytohormone production, and disease suppression, even in the absence of traditional soil environments.

Scientific Report

Plant Beneficial Bacteria and Their Potential Applications in Vertical Farming Systems
Country: United States
Publication Date: 15 January 2023
Main focus: This literature review explores the roles and applications of plant beneficial bacteria in vertical farming systems, emphasising their impact on plant nutrition, defence mechanisms, and hormonal regulation within controlled environment agriculture.

Key findings:
- Plant growth-promoting bacteria (PGPB) can significantly enhance nutrient uptake and plant growth, with notable improvements observed in macronutrient absorption.
- For instance, inoculation with *Rhizobia spp.* in hydroponic beans maintained normal nitrogen levels without chemical fertilizers.
- The global biofertilizer market is dominated by nitrogen-fixing bacteria, holding 79% market share.
- Studies show that hydroponic systems using PGPB can achieve up to 90% water savings and 85% fertilizer savings compared to traditional farming methods.

Reference: Chiaranunt P, White JF. Plant Beneficial Bacteria and Their Potential Applications in Vertical Farming Systems. *Plants* **2023** Jan 15;12(2):400. DOI: 10.3390/plants12020400.

Scientific Report

Vertical Farming: The Only Way Is Up?
Country: Belgium
Date of publication: December 21, 2021
Main focus: The development and sustainability of vertical farming systems, with a focus on technological advancements and the resilience of hydroponic cultivation.

Key findings: The study highlights the historical evolution of vertical farming, the integration of plant growth-promoting rhizobacteria (PGPR) to improve plant resilience and sustainability, and the economic, environmental, social, and political dimensions of vertical farming. It emphasises the importance of circularity in vertical farming systems, particularly through closed-loop hydroponics.

Reference: Van Gerrewey, T.; Boon, N.; Geelen, D. Vertical Farming: The Only Way Is Up? *Agronomy* **2022**, *12*, 2. DOI: 10.3390/agronomy12010002

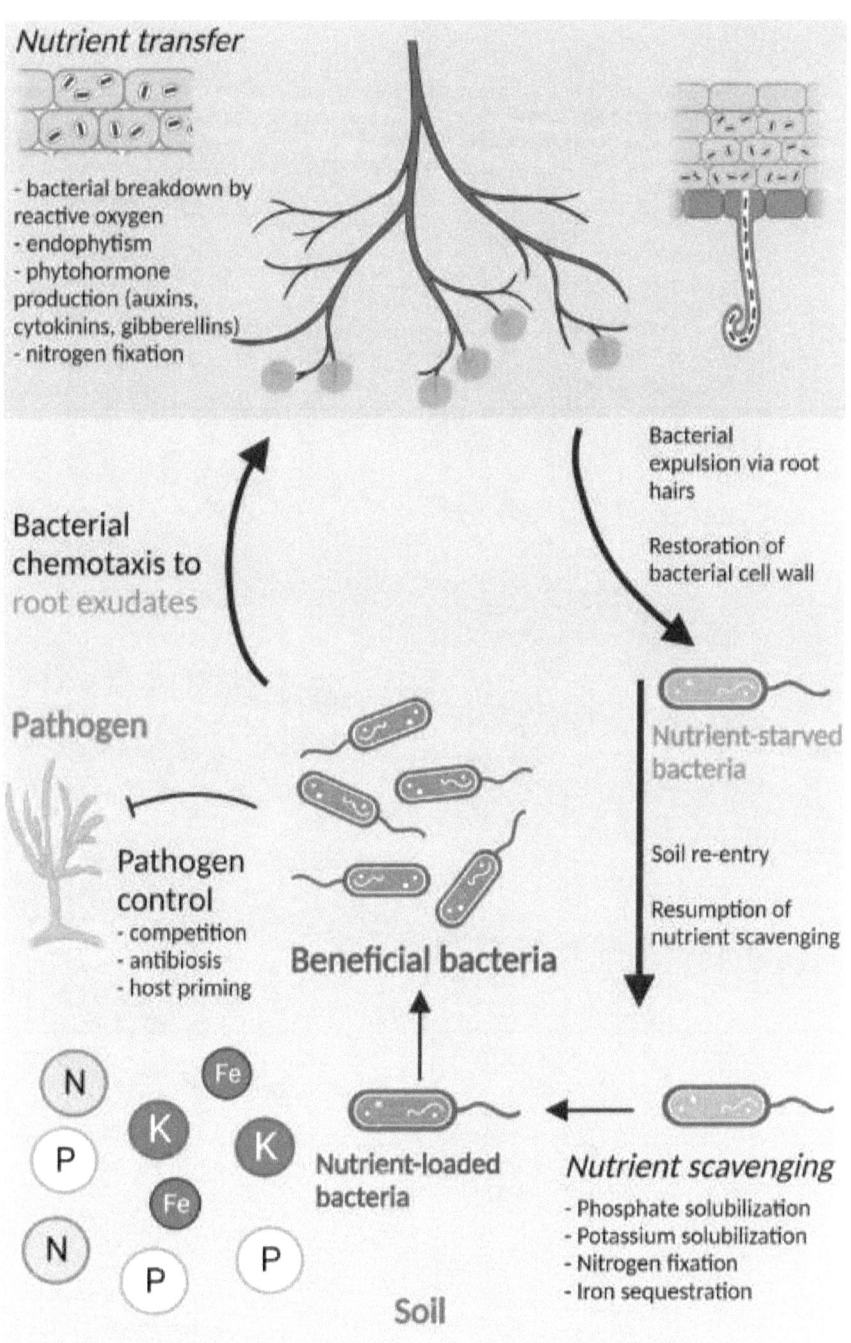

A schematic representation of beneficial bacteria's growth-promotional and defensive functions, including their role in the rhizophagy cycle. The host plant (green) breaks down soil bacteria with ROS, enabling endophytism and nutrient transfer. Nutrient-starved bacteria are expelled via root hairs, restore their cell walls in soil, and resume nutrient scavenging, including phosphate and potassium solubilization, nitrogen fixation, and iron sequestration. Nutrient-loaded bacteria (blue) are attracted back to the plant by root exudates, where they are degraded by ROS for nutrient transfer. Beneficial bacteria also help control pathogens through competition, antibiosis, and priming of plant resistance. Source: Chiaranunt & White , 2023

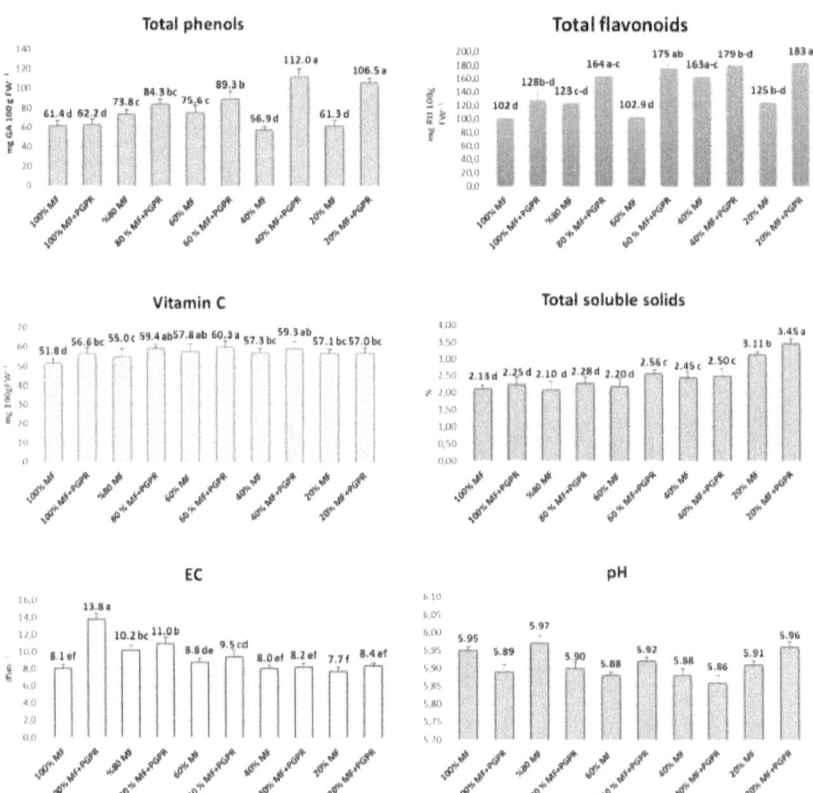

Effects of PGPR on total phenols, total flavonoids, vitamin C, TSS, EC and pH of lettuce. MF Mineral fertiliser, PGPR Plant growth promoting Rhizobacteria. There is no significant difference between means with the same letter in the same histogram (p < 0.05).
Source: Ikiz et al., 2024

The layout of the experiment in the floating culture hydroponic system; growing lettuce with PGPR using reduced mineral fertilizer. Source: Ikiz et al., 2024

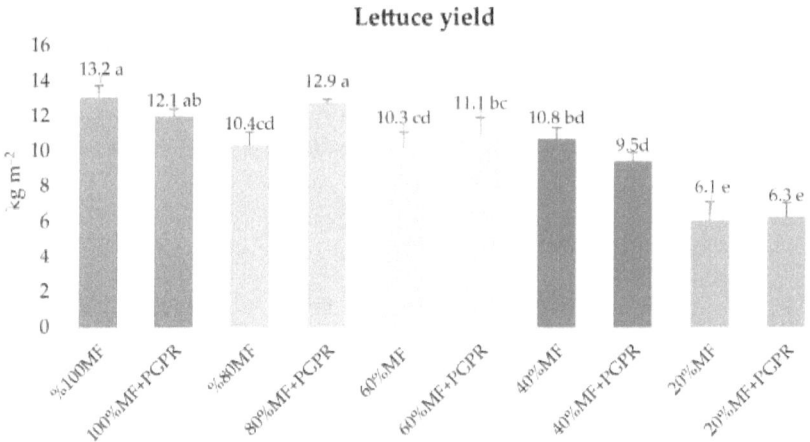

Effects of the PGPR on lettuce yield under the reduced mineral fertilizer levels. MF Mineral fertilizer, PGPR Plant growth promoting Rhizobacteria. There is no significant difference between means with the same letter in the same histogram ($p < 0.05$). Source: Ikiz et al., 2024

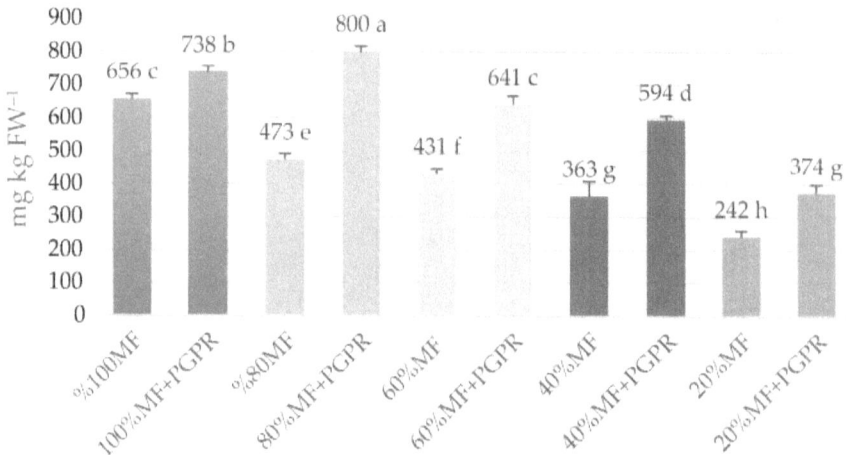

Effects of PGPR on hydroponically grown lettuce nitrate content under the reduced mineral fertilizers. MF Mineral fertiliser, PGPR Plant growth promoting Rhizobacteria. There is no significant difference between means with the same letter in the same histogram ($p < 0.05$). Source: Ikiz et al., 2024

Scientific Report

The Role of Plant Growth-Promoting Microorganisms (PGPMs) and Their Feasibility in Hydroponics and Vertical Farming

Country: Saudi Arabia
Publication Date: 9 February 2023
Main focus: The study reviews the application and benefits of plant growth-promoting microorganisms (PGPMs) in hydroponic and vertical farming systems, highlighting their potential to enhance plant growth and stress tolerance.

Key findings:
- The global market for vertical farming is predicted to exceed USD 10.02 billion by 2027.
- Soil-based agriculture consumes 20 times more water compared to hydroponics. The incorporation of PGPMs in hydroponic systems can enhance nutrient uptake and plant growth.
- For instance, *Gluconacetobacter diazotrophicus* and *Azospirillum brasilense* increased iron nutrition in strawberries.
- PGPMs can improve abiotic stress tolerance, such as drought and salinity, by regulating phytohormones and antioxidants.

- The use of PGPMs in hydroponics can reduce the cost of water treatment and pathogen control, thereby making these systems more sustainable and efficient.

Reference: Dhawi, F. The Role of Plant Growth-Promoting Microorganisms (PGPMs) and Their Feasibility in Hydroponics and Vertical Farming. *Metabolites* **2023**, *13*, 247. DOI: 10.3390/metabo13020247

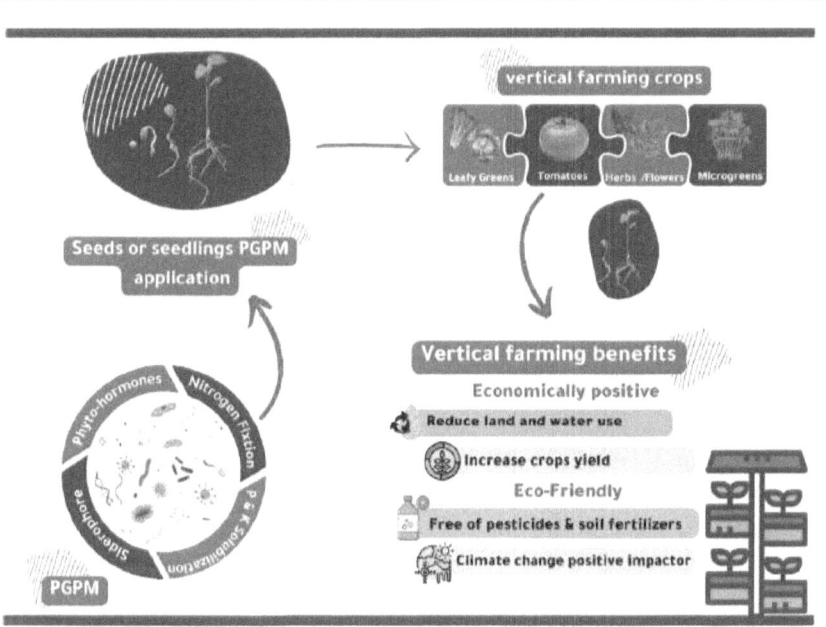

The types of plants used and the economic and environmental feasibility of growth promoting microorganisms in hydroponics and vertical farming. Source: Dhawi, 2022

Previous studies on plant growth-promoting microorganisms' (PGPMs) mode of application in a soil-free system, plant types used, and their influence. Source: Dhawi, 2022

Plant	Mode of PGPM Application	Type of PGPM	Influence
Banana Berangan' (*Musa* spp. dessert type)	Seedling inoculation with microbial suspension	*Bacillus sphaericus* and *azospirillum*	Increase in root formation, leaf area, chlorophyll content, and consequently, total biomass

Plant	Mode of PGPM Application	Type of PGPM	Influence
Strawberries	Siderophores added to hydroponic medium	*Gluconacetobacter diazotrophicus* and *azospirillum brasilense*	Increased the nutrition of iron
Triticum aestivum	Seedling inoculation with microbial suspension	*Calothrix sp., anabaena cylindrica, chryseobacterium balustinum, pseudomonas simiae,* and *pseudomonas fluorescen*	Increased the growth, plant height, dry shoot mass, total nutrients, and the ability to produce indole acetic acid
Lettuce	Biostimulant extract added to hydroponic medium	*Bacillus* spp.	Minimized salt stress
Tomato	Seedling inoculation with microbial suspension	*Penicillium brevicompactum, penicillium solitum strain 1, pseudomonas fluorescens subgroup g strain 2, pseudomonas marginalis, pseudomonas putida subgroup b strain 1, pseudomonas syringae strain 1,* and *trichoderma atroviride*	Plant growth and development in the absence of pathogens (antagonistic activity against *Pythium ultimum*)
Glycine max (L.) Merr.	Seedling inoculation with microbial suspension	Bacteria, yeasts, mycorrhiza, and Trichoderma	Higher density of smaller stomata, thicker palisade parenchyma, larger intercellular spaces in the mesophyll; increased photosynthetic traits, growth and seed production
Lettuce (Salanova® *Lactuca sativa* and Salanova® Red Crisp).	Bio-stimulant extract added to hydroponic medium	Phycocyanin-rich spirulina extract	Reduced time from seed to harvest by 6 days, increased yield by 12.5%, and improved antioxidant flavonoid levels

Artificial Pollination

For fruiting crops, **pollination** becomes another arena for innovation. Some vertical farms use mechanical vibration to simulate the action of bees, while others introduce actual bees into their enclosed systems, creating a harmonious blend of natural processes and high-tech agriculture.

> Artificial pollination is the **manual transfer of pollen from the male parts of a plant to the female parts**, typically performed by humans or machines. This technique is used when natural pollination is insufficient or impossible, often due to environmental factors, lack of pollinators, or in controlled breeding programs to produce specific plant varieties.

In vertical farming systems, artificial pollination techniques have been developed to overcome the challenges posed by the **absence of natural pollinators**.

Some farms introduce **natural pollinators**. For example, hoverflies are good pollinators. The larvae of many species feed on aphids, providing natural pest control, while others efficiently recycle nutrients, contributing to a balanced and sustainable growing environment.

Hoverflies on the feeder (left) and foraging a flower (right) on DAT 5.
Source: Carpineti *et al.*, 2024

Some farms employ robotic systems equipped with sensors and precision tools to identify flowers at the optimal stage for pollination and transfer pollen. These robots can work around the clock, ensuring consistent pollination rates across multiple levels of the vertical farm. Other operations utilise **targeted air circulation systems** that gently disperse pollen throughout the growing area, mimicking wind pollination in a controlled environment. These artificial pollination methods not only increase crop yields but also **allow for greater control over breeding and genetic diversity** in vertical farm produce.

Scientific Report

Precision Pollination Strategies for Advancing Horticultural Tomato Crop Production
Country: Australia 🇦🇺
Publication Date: 18 February 2022
Main focus: This review explores various precision pollination technologies and their potential to enhance tomato crop production in protected-cropping environments.

Key findings: Precision pollination using technologies such as robotic air and acoustic devices can significantly improve tomato seed set, size, yield, and quality. For example, bumblebee pollination resulted in a 74% increase in fruit weight compared to non-pollinated controls, whereas artificial pollination achieved a 28% increase. The review highlights the cost-effectiveness of bumblebees, which can pollinate up to 500 plants per day, compared to the estimated AUD 25,000/ha/year cost of manual pollination.

Reference: Dingley, A.; Anwar, S.; Kristiansen, P.; Warwick, N.W.M.; Wang, C.-H.; Sindel, B.M.; Cazzonelli, C.I. Precision Pollination Strategies for Advancing Horticultural Tomato Crop Production. *Agronomy* **2022**, *12*, 518. DOI: 10.3390/agronomy12020518

Relative costs of commercial pollination techniques in Australia.
Source: DiIngley *et al.*, 2022

Pollination Technique	Cost to Industry
Insect pollination	Moderate (~AUD 9300/ha/year)
Electric toothbrush	High (~AUD 25,000/ha/year)
Buzz probe	High (~AUD 25,000/ha/year)
Trellis tapping	Moderate (~AUD 10,764/ha/year)

The degree of human involvement required to operate the pollination technique is strongly linked to cost, with more involvement increasing cost. Previous costs associated with insect pollination, electric toothbrush and buzz probe, and trellis trapping (AUD 26/h: 3 h × 3 days/week × 46 weeks) are estimates only for comparison purposes.

Advantages and disadvantages of the different techniques available to pollinate tomato flowers in the protected-cropping industry. Source: DiIngley *et al.*, 2022

Pollination Technique	Advantages	Disadvantages
Insect buzz	High efficacy, low labour requirement	No bumblebees in Australia, rely upon native bees
Tuning fork or probe wand	High efficacy, can pollinate neighbouring floral truss effectively	High labour cost, restricted to only trusses near the placement of the probe
Trellis tapping	Moderate efficacy, low labour requirement, pollinates multiple trusses quickly	Moderate labour cost, may not pollinate all trusses, requires multiple tapping per week
Air jet pulse or wind movement	Moderate/low efficacy, prevision for precision robotic automation	Requires moderate capital investment, may contribute to foliar pathogen spread
Acoustic waves	Moderate/high efficacy, prevision for precision robotic automation	Not yet proven, requires capital investment, possible health and safety concerns due to noise.

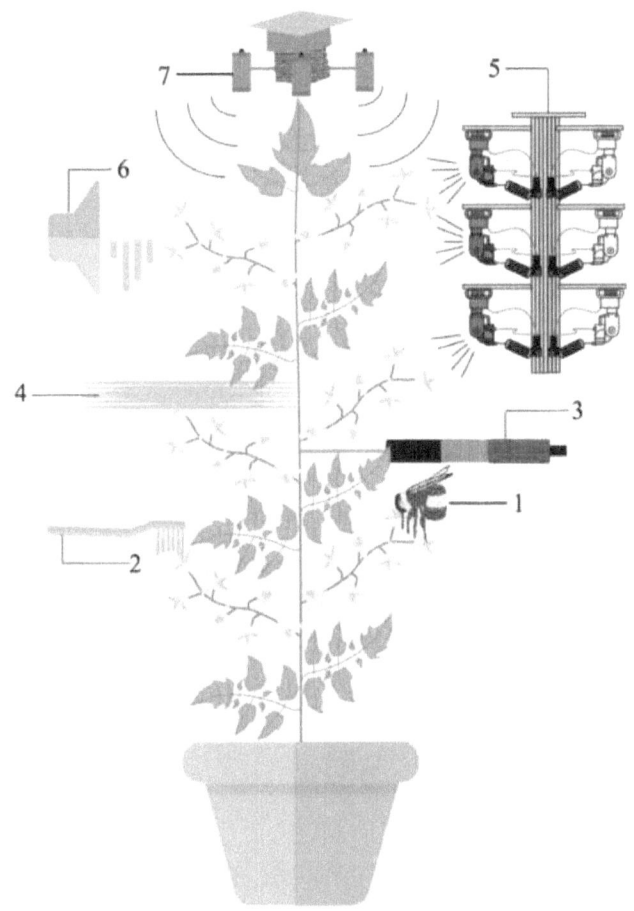

Stylised image showing different pollination techniques utilised in the tomato protected-cropping industry:
(1) Natural buzz by bees can effectively pollinate multiple tomato flowers within a protected environment;
(2) The electric toothbrush can mechanically stimulate individual flowers making it effective for cross-pollination breeding programs;
(3) A tuning fork or wand placed below the floral truss can vibrate the entire truss leading to efficient pollination;
(4) Wooden sticks tapped onto trellis strings multiple times will force plant vibrations leading to effective pollination of many trusses and flowers;
(5) Commercial robotic devices have been engineered to eject pulses of air that mechanically shake the floral truss leading to targeted pollination of flowers in a high throughput manner;
(6) Acoustic speakers could be commercially developed to emit sonic waves that accelerate floral movement without physical contact and trigger successful pollination of an entire floral truss;
(7) Robotic drone technology has been developed to force mechanical air movement of specific intensity above the crop leading to pollination of upper floral trusses.
Source: Dilngley *et al.*, 2022

7. Sustainability and Resource Management

This chapter, *Sustainability and Resource Management*, delves into the critical aspects of environmental stewardship and efficient resource utilization in vertical farming systems.

Integrated Rainwater Harvesting explores innovative techniques for **capturing and utilizing rainwater** in vertical farming operations.

To address the energy-intensive nature of vertical farming, *Renewable Energy Integration* examines strategies for **incorporating sustainable energy sources** through agrivoltaics practise into vertical farming systems.

The chapter concludes with *Life Cycle Assessment*, which provides a **comprehensive evaluation of the environmental impact** of vertical farming throughout its entire lifespan.

Integrated Rainwater Harvesting

Integrated Rainwater Harvesting is an innovative approach that blends water conservation with vertical farming systems.

> Integrated Rainwater Harvesting involves **collecting and storing rainwater** from rooftops, facades, and other surfaces of buildings where vertical farms are installed.

The harvested water is then filtered and used to irrigate the crops, creating a sustainable water cycle within the urban farming ecosystem. By incorporating rainwater harvesting, vertical farms can significantly reduce their reliance on municipal water supplies, lowering both water costs and environmental impact.

This integration is particularly beneficial **in regions with adequate rainfall**, where it can provide a substantial portion of the farm's water needs.

The system typically includes collection surfaces, gutters, storage tanks, and filtration mechanisms, all designed to maximise water capture and quality.

Rainwater harvesting in vertical farms helps **save water and reduce city flooding**. It's one of the ways to grow food while also helping with water management, making urban farming more environmentally friendly and useful for the community.

Scientific Report

A Long-Term Analysis of the Possibility of Water Recovery for Hydroponic Lettuce Irrigation in an Indoor Vertical Farm. Part 2: Rainwater Harvesting
Country: Poland
Publication Date: 30 December 2020
Main focus: The study evaluates the feasibility of using a rainwater harvesting system to meet the water needs for hydroponic lettuce cultivation in an indoor vertical farm in Wrocław, Poland.

Key findings: The proposed rainwater harvesting system could cover an average of 35.9% of the water needs for the vertical farm, saving approximately 146,510 litres of water annually. When combined with water recovery from exhaust air, the system's efficiency increases, covering up to 90.4% of water demand and saving 340,300 litres per year.

Reference: Jurga, A.; Pacak, A.; Pandelidis, D.; Kaźmierczak, B. A Long-Term Analysis of the Possibility of Water Recovery for Hydroponic Lettuce Irrigation in an Indoor Vertical Farm. Part 2: Rainwater Harvesting. *Appl. Sci.* **2021**, *11*, 310. DOI: 10.3390/app11010310

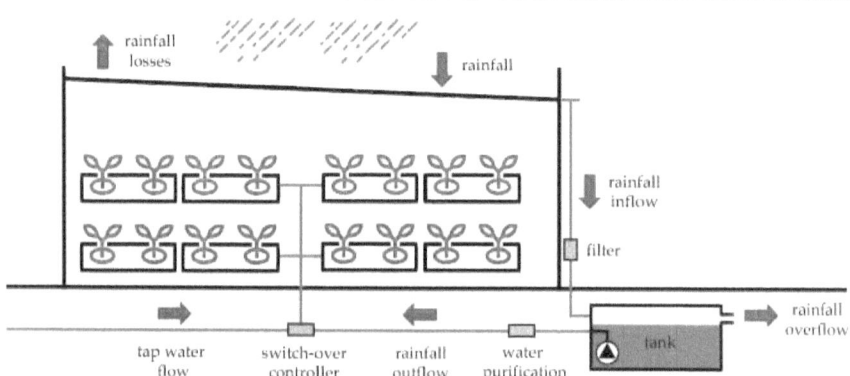

The diagram of the water flow in the rainwater harvesting system. Source: Jurga *et al.*, 2021

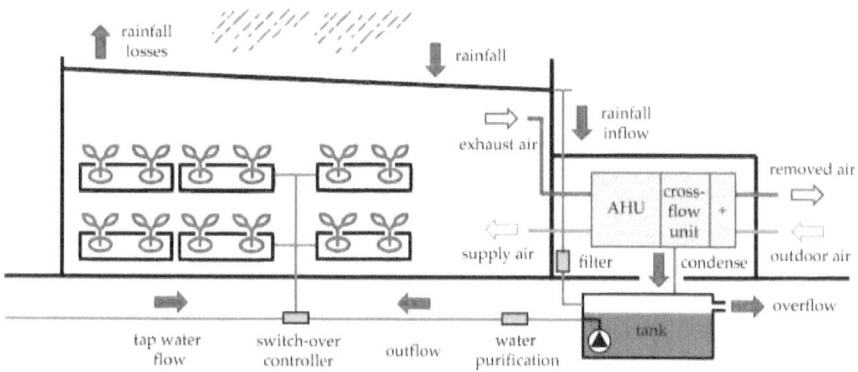

The diagram of the water flow in the rainwater harvesting system cooperating with water recovery from the exhaust air. Source: Jurga *et al.*, 2021

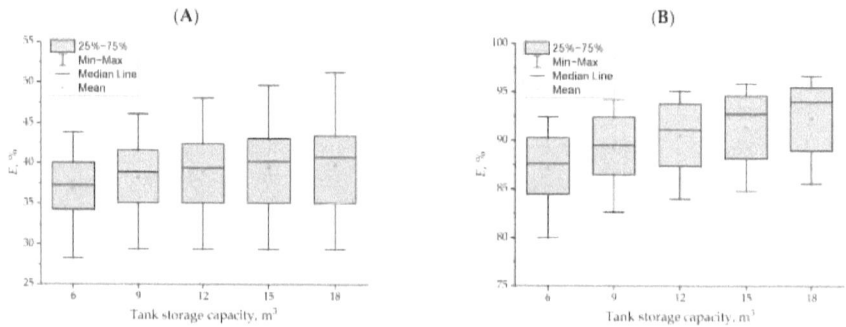

Comparison of the water-saving efficiency with different tank storage capacity for the rainwater harvesting (RWH) system (A), and the RWH system cooperating with water recovery from the exhaust air (B). Source: Jurga et al., 2021

Net present value (*NPV*) and payback period (*PBP*) values related to each tank size for the RWH system. Source: Jurga et al., 2021

Tank Volume, m3	Costs, €	Water-Saving, m3/Year	Annual E, %	NPV (20 Years), €	PBP, Year
6	2654	106 (min.)	28.3	1150	16
		138 (av.)	**36.8**	**2292**	**13**
		165 (max.)	43.9	3240	11
9	2863	111 (min.)	29.4	1092	16
		144 (av.)	**38.3**	**2279**	**13**
		173 (max.)	46.1	3335	12
12	3017	111 (min.)	29.4	938	17
		147 (av.)	**38.9**	**2219**	**14**
		181 (max.)	48.0	3438	12
15	3362	111 (min.)	29.4	593	18
		149 (av.)	**39.5**	**1945**	**15**
		187 (max.)	49.6	3308	12
18	3516	111 (min.)	29.4	439	19
		150 (av.)	**39.8**	**1841**	**15**
		193 (max.)	51.2	3368	13

Renewable Energy Integration

The rising costs of fossil fuels and electricity have highlighted the **need for energy-saving strategies and alternative energy sources** for vertical farming operations.

The primary renewable energy sources (RES) utilised in vertical farms include **solar panels, wind turbines, hydroelectric power, biomass, biofuels,** and **geothermal energy**. These systems support the use of electricity-based technologies, such as heat pumps, which consume significantly less energy than traditional fuel-fired units and reduce carbon dioxide emissions by 56–79% [Vatistas et al., 2022].

Some of the vertical farms have also incorporated **integrated photovoltaic (PV) panels** to produce electricity, meeting part of the energy demand and potentially selling excess electricity back to the grid.

Additionally, **battery installations** can store electricity generated by PV panels or other RES, providing power during periods of insufficient sunshine or other limiting conditions.

An emerging approach in the renewable energy integration field is **agrivoltaics**. Agrivoltaics combines **solar panels with farming** by placing panels above crops, allowing the same land to produce both renewable energy and support agricultural production simultaneously.

In vertical farming contexts, this might involve **using semi-transparent solar panels** as part of the building envelope, allowing **some light through for plant growth** while generating electricity. This dual-use approach can potentially increase land-use efficiency and provide additional income streams for vertical farming operations.

Scientific Report

Increasing the agricultural sustainability of closed agrivoltaic systems with the integration of vertical farming: A case study on baby-leaf lettuce
Country: Italy
Publication Date: 24 May 2023

Main focus: This study investigates the integration of vertical farming into closed agrivoltaic systems to enhance the sustainability and productivity of agricultural practices, specifically focusing on baby-leaf lettuce.

Key findings: The integration increased yield by 13 times compared to conventional agrivoltaic systems. The system's average Land Equivalent Ratio (LER) was 1.31, but only 12% of the energy consumption was covered by the agrivoltaic energy, requiring additional PV panels to achieve energy self-sufficiency. The land consumption for PV panels was found to be 5 to 14 times the area of the vertical farm.

Reference: Cossu, M., Tiloca, M. T., Cossu, A., Deligios, P. A., Pala, T., & Ledda, L. Increasing the agricultural sustainability of closed agrivoltaic systems with the integration of vertical farming: A case study on baby-leaf lettuce. *Applied Energy* **2023**, Volume 344, 121278. DOI: 10.1016/j.apenergy.2023.121278

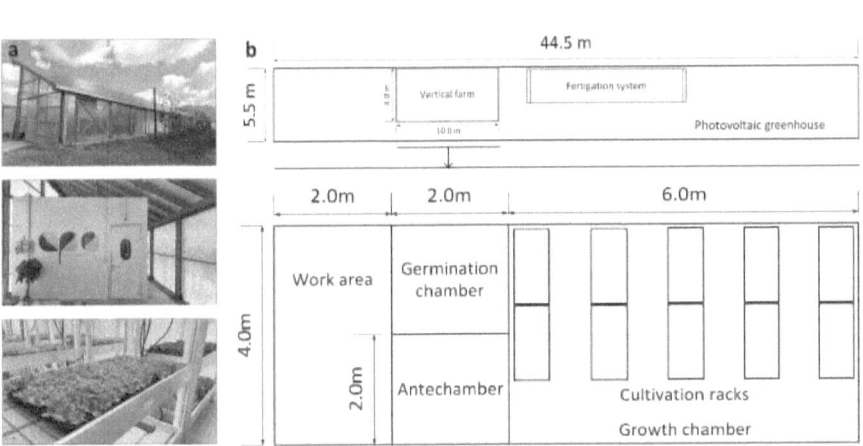

Experimental closed agrivoltaic system with integrated vertical farm. External and internal view of the VF (a) and general layout (b). Source: Cossu *et al.*, 2023

Romaine lettuce control cycle inside a closed agrivoltaic system module identical and close to the vertical farm integrated in a closed agrivoltaic system. The photoradiometers are located in the centre of each row. Source: Cossu et al., 2023

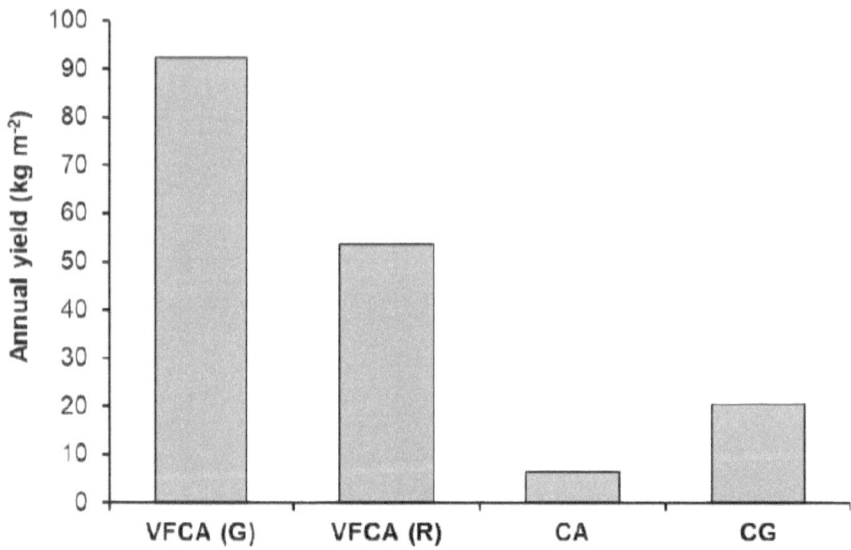
Estimated annual yield of the vertical farm integrated in a closed agrivoltaic system with green (G) and red (R) varieties, in comparison to the control closed agrivoltaic system and a conventional greenhouse (CG). Source: Cossu et al., 2023

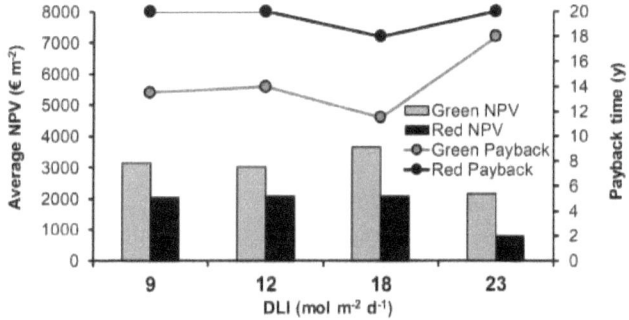

Average net present value (NPV) over 20 years and payback time of the vertical farm integrated in a closed agrivoltaic system for the green and red varieties as a function of the DLI. Source: Cossu et al., 2023

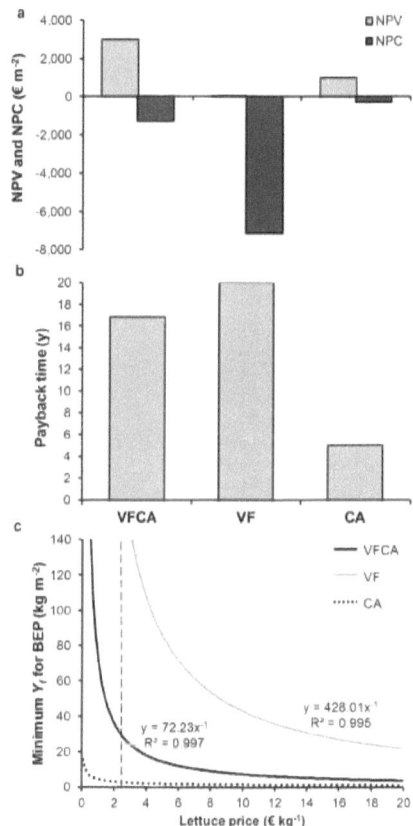

Average net present value (NPV) over 20 years and average payback time of the vertical farm integrated in a closed agrivoltaic system (a), average payback time (b) and minimum yield necessary to reach the breakeven point, of the vertical farm integrated in a closed agrivoltaic system, the VF and the control closed agrivoltaic system (c), where the average baby leaf lettuce price of 2.5 € kg−1 is highlighted with a vertical dotted line. Source: Cossu et al., 2023

Scientific Report

Lettuce Production under Mini-PV Modules Arranged in Patterned Designs
Country: Spain
Publication Date: 15 December 2021
Main focus: The study investigates the impact of different shading treatments using photovoltaic (PV) modules on the growth and yield of lettuce grown on rooftops.

Key findings: The research found that lettuce grown under scattered shade (SS) treatment produced 46.4% more fresh weight in spring and 61.2% more in summer compared to the concentrated shade (CS) treatment. Additionally, SS treatment resulted in a 68.8% higher yield in spring and 87.6% higher in summer compared to full sun (FS) treatment.

Reference: Carreño-Ortega, A.; do Paço, T.A.; Díaz-Pérez, M.; Gómez-Galán, M. Lettuce Production under Mini-PV Modules Arranged in Patterned Designs. *Agronomy* **2021**, *11*, 2554. DOI: 10.3390/agronomy11122554

Experimental design. SS (Scattered shade), CS (Concentrated shade), FS (Full Sun), R (treatment replication), BPN (Bird proof net), PAR (quantum PAR sensor, within a discontinuous line indicated below the panel), TM (Thermometer).
Source: Carreño-Ortega *et al.*, 2021

Aerial photograph of the experimental design. Source: Carreño-Ortega *et al.*, 2021

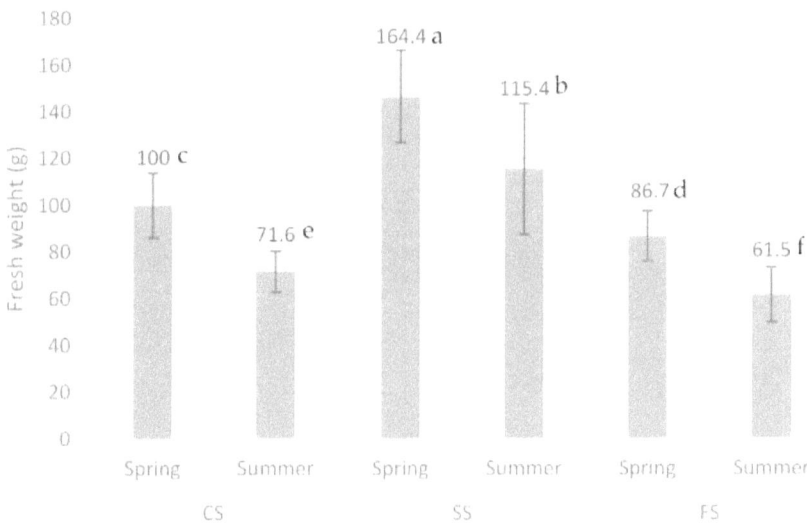

Average lettuce fresh weight (g) and errors, during spring and summer season. Different letters indicate significance differences, with ANOVA significance level α = 0.05. (CS) Concentrated shadow treatment. (SS) Scattered shadow treatment. (FS) Full sun treatment. Source: Carreño-Ortega *et al.*, 2021

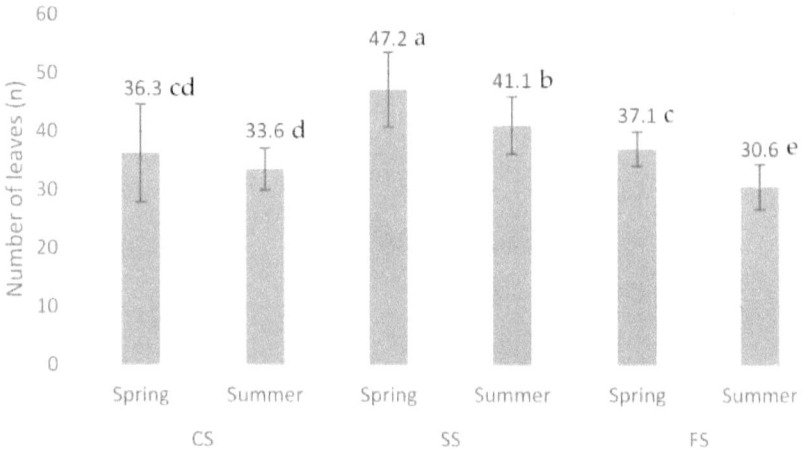

Average lettuce number of leaves (n) and errors during spring and summer season. Different letters indicate significance differences, with ANOVA significance level α = 0.05. (CS) Concentrated shadow treatment. (SS) Scattered shadow treatment. (FS) Full sun treatment. Source: Carreño-Ortega *et al.*, 2021

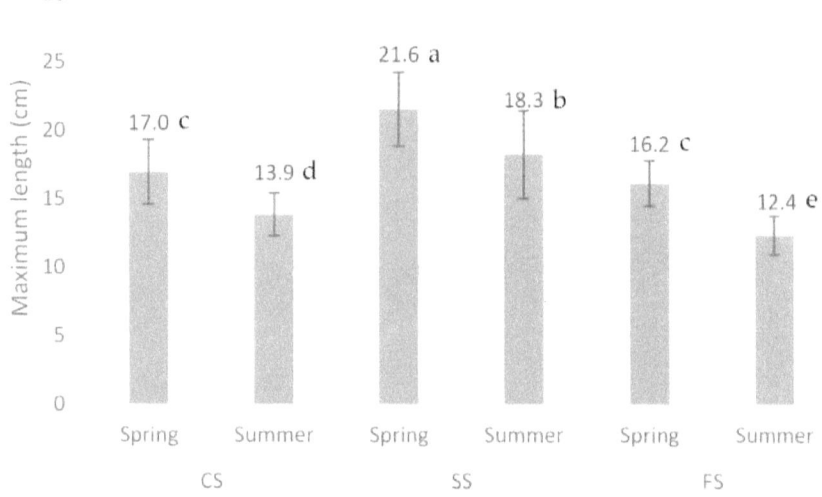

Average lettuce maximum length of the leaves (cm) and errors, during spring and summer seasons. Different letters indicate significance differences, with ANOVA significance level α = 0.05. (CS) Concentrated shadow treatment. (SS) Scattered shadow treatment. (FS) Full sun treatment. Source: Carreño-Ortega *et al.*, 2021

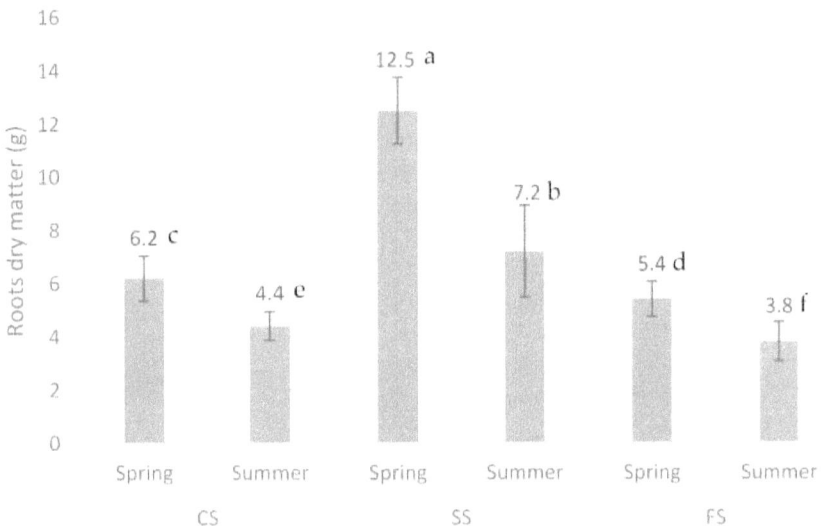

Average lettuce roots dry matter (g) and errors, during spring and summer seasons. Different letters indicate significance differences, with ANOVA significance level α = 0.05. (CS) Concentrated shadow treatment. (SS) Scattered shadow treatment. (FS) Full sun treatment.
Source: Carreño-Ortega et al., 2021

Life Cycle Assessment

Life cycle assessment (LCA) is a key method for **evaluating the environmental impact** of vertical farms from setup to operation and decommissioning. This approach offers valuable insights into the sustainability of these farming systems.

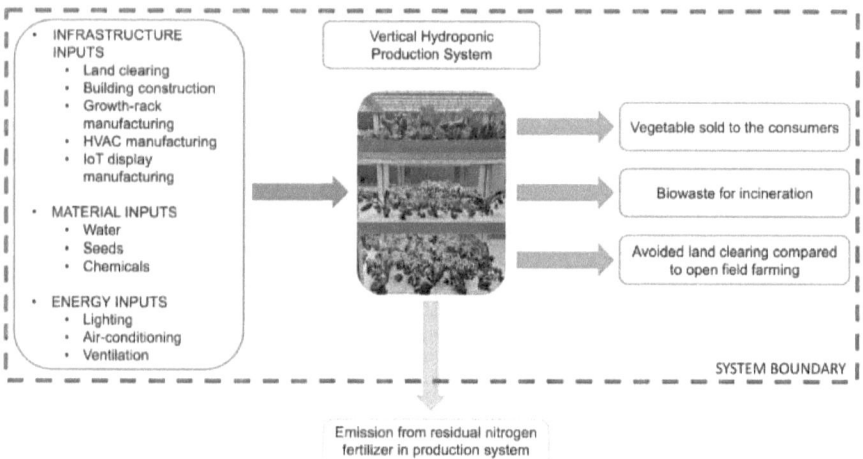

System boundary for the life-cycle-based carbon footprint analysis. Source: Song *et al.*, 2024

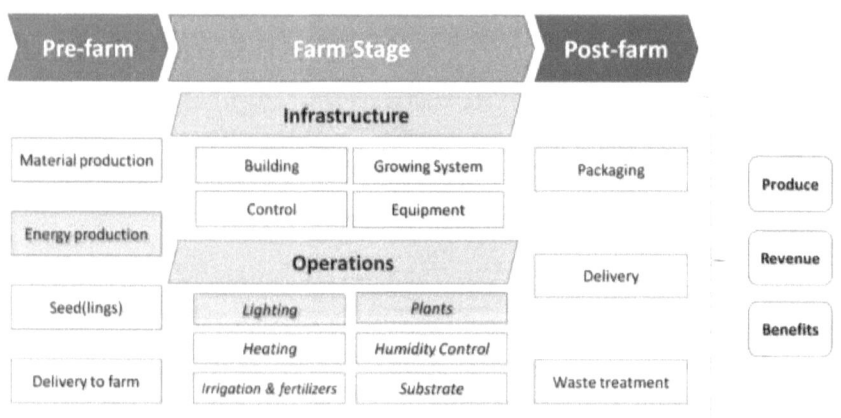

Vertical farming system activities organised in a life cycle assessment process diagram to highlight the carbon costs in controlled environment agriculture. Groups highlighted in red represent activities with high environmental impact that are addressed in this review. Source: de Carbonnel *et al.*, 2022

Research has shown that the **carbon footprint of vertical farms can be much higher** than traditional field-grown crops when using conventional energy sources. For example, a study by [Beacham et al., 2019] found that CO2 emissions per kilogram of lettuce from vertical farms were five times higher than field-grown lettuce in summer and twice as high in winter.

However, this presents **an opportunity for improvement**.

The high energy use in vertical farming, mainly due to artificial lighting and climate control, highlights the **need for renewable energy integration**. Using solar, wind, or other sustainable energy sources can significantly reduce the carbon footprint of vertical farms.

Additionally, vertical farming often uses **less water and fewer pesticides** than traditional agriculture. Producing food **closer to urban centres** can also cut down transportation emissions. When considering these factors in an LCA, **the overall environmental impact of vertical farms can be more favourable.**

Scientific Report

Does Green Vertical Farming Offer a Sustainable Alternative to Conventional Methods of Production?: A Case Study from Scotland
Country: United Kingdom
Publication Date: 5 October 2022
Main focus: The study evaluates the environmental impact of vertical farming (VF) compared to conventional open-field and greenhouse farming methods in Scotland, with a focus on carbon footprint, water usage, and energy consumption.

Key findings: The research indicates that electricity consumption accounts for 91% of the carbon footprint in vertical farming. Under Scotland's 2020 electricity mix, the carbon emissions for VF-produced lettuce were approximately 0.42 kg CO2-eq per kg, which is comparable to the emissions from UK open-field agriculture (0.46 kg CO2-eq per kg). The potential for further reductions to 0.33 kg CO2-eq per kg exists under a 100% renewable electricity scenario.

Reference: Sandison, F., Yeluripati, J., & Stewart, D. (2023). Does green vertical farming offer a sustainable alternative to conventional methods of production?: A case study from Scotland. *Food and Energy Security* **2023**, 12, e438. DOI: 10.1002/fes3.438

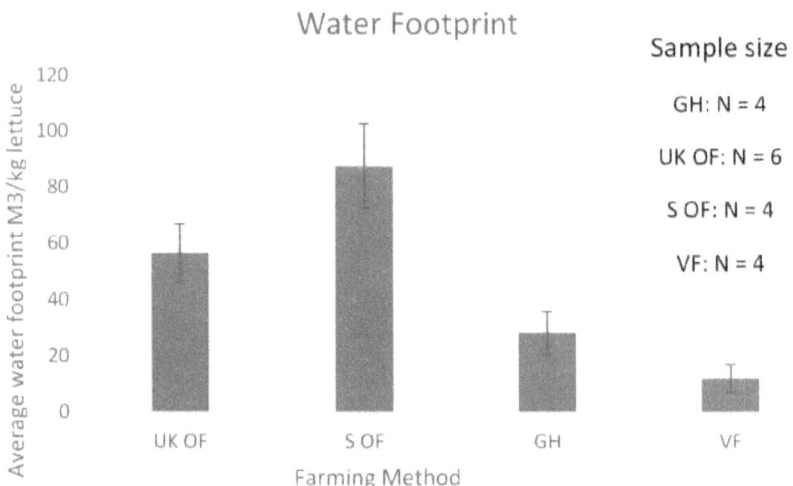

Comparison of water usage scores for selected farming methods found in literature where UK OF, open farm in the UK; S OF, open farm in Spain; GH, glasshouse; VF, vertical farm. Bars indicate statistical error. Source: Sandison et al., 2023

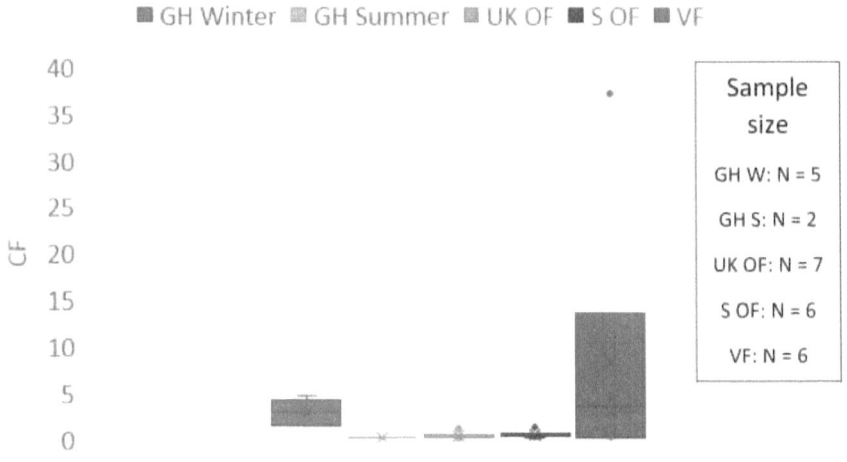

Comparative carbon footprint ranges (kg CO_2 eq.) found in literature for producing 1 kg lettuce, where GH W, glasshouse grown in winter; GH S, glasshouse grown in summer; UK OF, open farm grown in the UK; S OF, open farm grown in Spain; VF, vertical farm grown. Whiskers on plot represent the range, X the average value, and dots representative of outliers. Source: Sandison et al., 2023

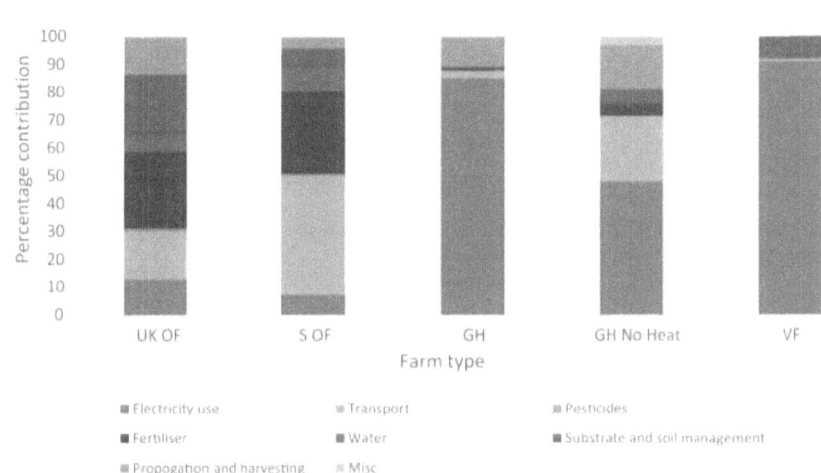

Analysis of relative contributions towards overall carbon footprint score for selected farming methods, where UK OF, open farm in the UK; S OF, open farm in Spain; GH, heated glasshouse; GH no heat, unheated glasshouse; VF, vertical farm.
Source: Sandison et al., 2023

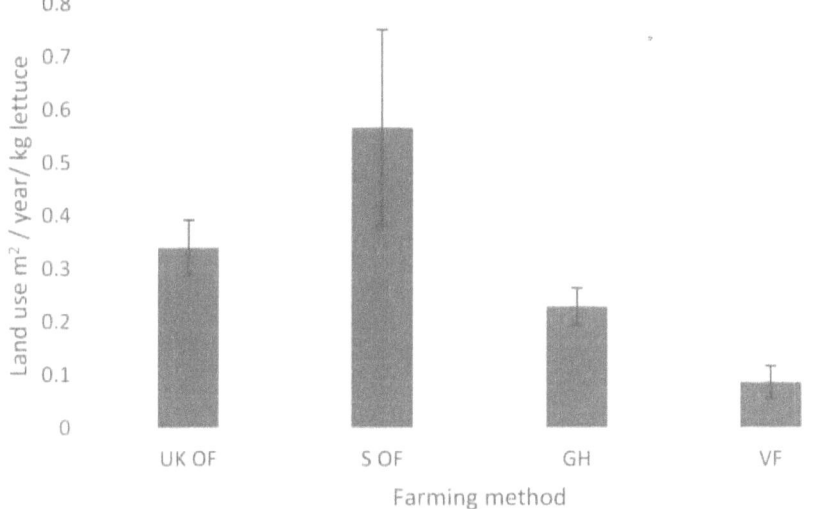

Comparison of land usage scores for selected farming methods found in literature where UK OF, open farm in the UK; S OF, open farm in Spain; GH, glasshouse; VF, vertical farm. Bars indicate statistical error. Source: Sandison et al., 2023

Scientific Report

The role of aeroponic container farms in sustainable food systems – The environmental credentials

Country: United Kingdom
Publication Date: 23 November 2022
Main focus: This study assesses the environmental impacts of aeroponic container farms in the UK, examining their potential to contribute to sustainable urban food production and reduce climate impacts.

Key findings:
- The aeroponic container farm system generates 1.52 kg CO2eq. per kg of pea shoots using the 2021 UK grid.
- Utilising solar and wind power can lower greenhouse gas emissions of aeroponic container farms by up to 80%.
- The energy source is critical in reducing most of the environmental impacts, making renewable-powered aeroponics more competitive with conventional agricultural methods.
- The study also indicates that aeroponic container farms can produce food with lower environmental impacts than equivalent imports to the UK.

Reference: Schmidt Rivera X, Rodgers B, Odanye T, Jalil-Vega F, Farmer J. The role of aeroponic container farms in sustainable food systems – The environmental credentials. *Science of The Total Environment.* **2023**. 860:160420. DOI: 10.1016/j.scitotenv.2022.160420

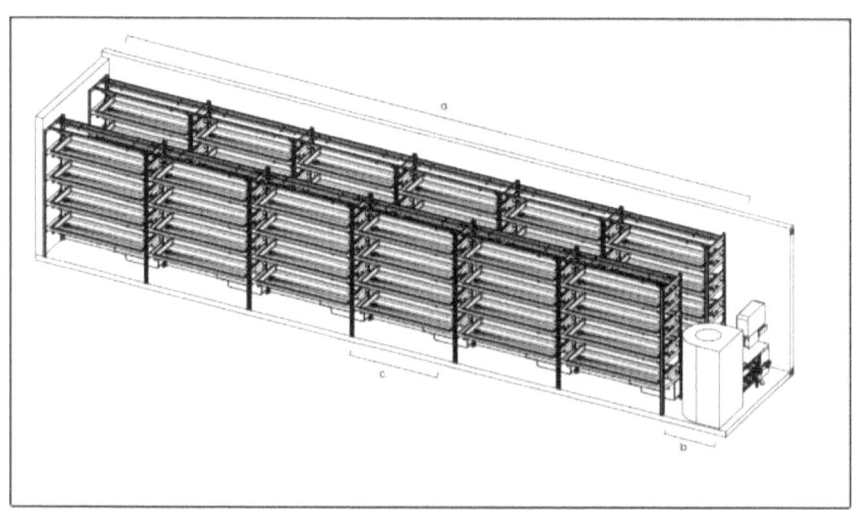

Angled view of the layout within the shipping container. a). the growing area, b). the water system comprising a reservoir, filter, and nutrient dosing system (water chiller).

Source: Schmidt Rivera et al., 2023

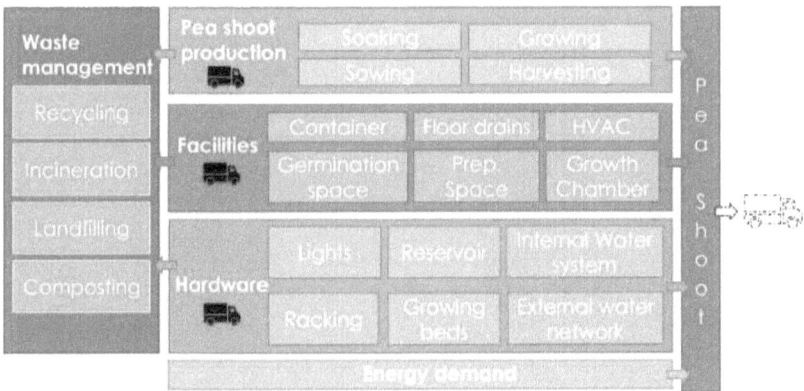

Life cycle stages of the aeroponic container system. Source: Schmidt Rivera et al., 2023

**Energy demand of the aeroponic container farm.
Source: Schmidt Rivera et al., 2023**

Activity	kWh/f.u.	Share
Bed controllers	0.63	13 %
Environment	1.06	22 %
Facilities	0.12	2 %
Fertigation	0.04	1 %
Irrigation	1.25	26 %
Lighting	1.69	34 %
Operations	0.11	2 %
Soaking	0.001	0.01 %
Total	4.9	100 %

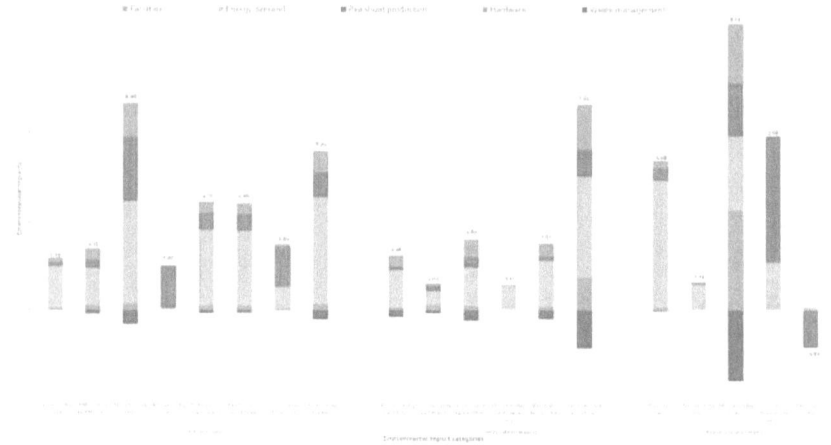

Environmental impacts of 1 kg of pea shoot produced in an aeroponic container farm system.
Source: Schmidt Rivera et al., 2023

**Analysis of the influence of different UK energy mixes.
Source: Schmidt Rivera et al., 2023**

Year	Carbon intensity [kg CO_2eq./kWh]	CC of aeroponic production [kg CO_2eq./fu]	Improvement
2017	0.384	2.28	This study
2018	0.307	1.89	17 %
2019	0.277	1.75	23 %
2020	0.253	1.63	28 %
2021	0.231	1.52	33 %
2024	0.111	0.93	59 %
2026	0.098	0.87	62 %
2028	0.100	0.88	61 %
2030	0.085	0.80	65 %
2032	0.064	0.70	69 %
2034	0.051	0.64	72 %
Green Tariff*	0	0.39	83 %

* For the calculations, 0 kg CO2 eq./kWh of energy acknowledging that this value might be under Scope 1 for reporting GHG emissions while the system accounts for scope 3.

8. Case Studies of Successful Vertical and Urban Farms

Vertical farming has seen numerous success stories around the world. These case studies demonstrate the variety and potential of urban agriculture. While avoiding purely commercial initiatives, below is a list of some fundamental cases and collaborations.

FAO Green Cities Initiative ●

This ambitious project by the Food and Agriculture Organization (FAO) supports urban and peri-urban agriculture globally. While not solely focused on vertical farming, it embraces innovative techniques to green cities: from rooftop gardens to community allotments, the initiative is transforming urban landscapes into productive green spaces. Several cities have been highlighted for their involvement.

> These cities are at **different stages of implementation**, with some just beginning to develop their plans and others already implementing specific projects in, primarily, outdoor vertical farming.

FAO is promoting vertical farming as one of the core components of urban and peri-urban agriculture. Source: *FAO* / Green Cities Initiative

Milan Urban Food Policy Pact 🏳

Supported by the FAO, this international agreement is a game-changer for urban agriculture and vertical farming. With **over 200 signatory cities**, it promotes **sustainable food systems**, including vertical farming projects.

Milan's Porta Nuova district showcases an **innovative blend of urban architecture and nature** with the Bosco Verticale, two residential towers featuring nearly 1,000 trees on their balconies. Designed by Stefano Boeri Architetti and completed in 2014, this vertical forest helps **mitigate air pollution** and **regulate building temperature**.

The development extends to include a **temporary wheat field art installation**, connecting the towers to nearby corporate skyscrapers.

Bird's-eye view of Bosco Verticale. Source: Stefano Boeri Architetti

Internal view of Bosco Verticale. Source: Stefano Boeri Architetti

Nairobi, Kenya Urban Farming Project

In Kenya's capital, Nairobi, the FAO has supported an urban agriculture initiative that has **repurposed vacant lots into productive farms using vertical farming techniques** to optimise space utilisation.

This initiative not only focuses on food production but also contributes to job creation and enhances food security for many Nairobi residents.

Kevin Abuya works on Victor Edalia's expanded urban farm, which now provides free vegetables to hundreds of families each month. Source: *NPR* / Thomas Bwire

Kigali, Rwanda, Simple Urban Hydroponics

A pilot project in Kigali, Rwanda, implemented from September 2017 to November 2019, provided subsidized hydroponics kits to low-income farmers. The initiative funded by USAID, aimed to introduce an alternative method for growing leafy greens and vegetables using minimal water and no soil.

Key achievements included distributing **41 hydroponics kits and reducing water usage by 30% compared to traditional irrigation practices**.

A general view of a vertical farming setup in the Simple Urban Hydroponics project.
Source: *VIA Water*

Quezon City, Philippines, Joy of Urban Farming

Quezon City is the largest city in Metro Manila. Its local government has implemented several successful programs, including the Joy of Urban Farming (JOUF). These initiatives focus on urban farming, utilizing both private and public spaces to improve food productivity and sovereignty. The JOUF program has **established 166 urban farms and 7 community model farms.**

Entrance to one of the "Joy of Urban Farming" settings.
Source: Global Covenant of Mayors for Climate & Energy

Dakar, Senegal Urban Horticulture Project

Another FAO-supported initiative, this project in Dakar showcases how **vertical farming can thrive in a West African context.**

By utilizing wall spaces and rooftops, the project has enabled urban dwellers to grow fresh produce year-round. It's a testament to how vertical farming can be adapted to different climates and cultures.

Micro gardening in Dakar alleviates poverty, hunger and food insecurity. Source: *C40*

Antananarivo, Madagascar, GRET Urban Farming,

In response to severe food security challenges in Antananarivo, Madagascar's capital, an urban agriculture project funded by Agence française de développement has been implemented since 2022.

The initiative focuses on developing **small-scale farming systems in available urban spaces**, prioritising short-cycle crops to improve nutrition for 800 vulnerable households, particularly women.

Urban farmer Bakoly Ralaiarisoa in her vegetable garden. Source: *GRET*

Lima, Peru, H2Grow

The World Food Programme (WFP) has implemented an **innovative hydroponics project called H2Grow** in Lima, Peru, to address malnutrition in urban slums.

The project teaches vulnerable communities, especially women, **how to grow nutritious food** using soilless cultivation techniques in desert-like conditions.

One of the H2Grow communities. Source: *WFP Innovation Accelerator*

Open Agriculture Initiative (OpenAg), USA

Based at MIT Media Lab in Cambridge, Massachusetts, USA, this project **ran from 2015 to 2020**. OpenAg aimed to create an open-source ecosystem for food production technologies, including vertical farming. Their **"Food Computers"** - controlled environment systems for crop experimentation - captured the imagination of researchers and hobbyists alike. Although the project has concluded, its open-source ethos continues to inspire innovation in urban agriculture.

May 2017 photograph of two food computers at the National Center for Agricultural Research and Extension facility at Al-Khalydeha Salinity Research Station in Mafraq, Jordan.
Source: *WFP Innovation Accelerator*

Scientific Report

Vertical Farm: from Agriculture to a New City Architecture
Country: Italy
Publication Date: October 2022
Main focus: This paper explores the intersection of high-rise construction and soilless agriculture, examining the potential of vertical farms to redefine urban architecture and agricultural practices.

Key findings:
- The study identifies three paradigmatic cases of vertical farms, including Richard Rogers' Skyfarm, Carlo Ratti's Jian Mu Tower, and Chris Precht's Farmhouse.
- Vertical farms offer significant land savings and reduced water consumption, with hydroponic systems using about 90% less water and aeroponic systems over 95% less.
- The potential food production of the Jian Mu Tower is estimated at 270 tons per year, sufficient to meet the needs of approximately 40,000 people.
- Vertical farms can lead to substantial reductions in supply chain logistics and energy use, promoting urban self-sufficiency and sustainability.

Reference: Bisiani, T., Basso, S., Martolana, P., & Venudo, A. (2022). Vertical Farm: from Agriculture to a New City Architecture. Forum A+P **2022**, October. University of Trieste. Trieste Press

Richard Rogers' Skyfarm. Source: Bisiani *et al.*, 2022 / Corriere Della Sera

What's next?

More people. Less land. Climate change. This is our future, and we cannot avoid it. Therefore, we need to accept and adapt our food system accordingly.

Vertical farming offers a **powerful tool** for **reimagining urban environments** and **making non-arable climates productive** to enhance food security. As renewable energy becomes more affordable, the cost of operating vertical farms is expected to decrease, making them more affordable and widespread.

In the future, we may see entire city blocks transformed into productive green spaces, with skyscrapers doubling as farms and rooftop gardens interconnected by aerial walkways.At the same time, our current goal is to **spread awareness and knowledge about vertical farming** and empower more people to adopt this method of food production. In addition to commercial growers, community-managed vertical farms could become social hubs, fostering local food cultures and education.

This vision of sustainable urban futures sees cities as complex, integrated ecosystems where food production is as much a part of urban infrastructure as roads or electricity grids. **Urban farms should be common in cities worldwide.**

We can bring this vision to life **together.**

Bibliography

1. Ahammed, G.J., Guang, Y., Yang, Y. et al. Mechanisms of elevated CO2-induced thermotolerance in plants: the role of phytohormones. *Plant Cell Rep* **2021**. 40, 2273–2286. https://doi.org/10.1007/s00299-021-02751-z
2. Agarwal, A., de Jesus Colwell, F., Bello Rodriguez, J. *et al.* Monitoring root rot in flat-leaf parsley via machine vision by unsupervised multivariate analysis of morphometric and spectral parameters. *Eur J Plant Pathol* **2024**, 169, 359–377. https://doi.org/10.1007/s10658-024-02834-z
3. Al Mamun, M. R., Deb, I., Hridoy, T., Soeb, M. J. A., & Shammi, S. Effects of Different Lighting Conditions on Growth, Yield and Nutrient Content of White Oyster Mushroom in Vertical Farm. *European Journal of Agriculture and Food Sciences* **2021**, 3(6), 61–67. https://doi.org/10.24018/ejfood.2021.3.6.418
4. Alneyadi, K.S.S., Almheiri, M.S.B., Tzortzakis, N., & Di Gioia, F. Organic-Based Nutrient Solutions for Sustainable Vegetable Production in a Zero-Runoff Soilless Growing System. *Journal of Agriculture and Food Research* **2024**, 15, 101035. https://doi.org/10.1016/j.jafr.2024.101035
5. Arcasi A, Mauro AW, Napoli G, Tariello F, Vanoli GP. Energy and cost analysis for a crop production in a vertical farm. *Applied Thermal Engineering*. **2024**. 239:122129. https://doi.org/10.1016/j.applthermaleng.2023.122129
6. Asseng, S., Guarin, J. R., Raman, M., & Gauthier, P. P. G. Wheat yield potential in controlled-environment vertical farms. *Proceedings of the National Academy of Sciences* **2020**, 117(32), 19131-19135. https://doi.org/10.1073/pnas.2002655117
7. Bafort, F.; Kohnen, S.; Maron, E.; Bouhadada, A.; Ancion, N.; Crutzen, N.; Jijakli, M.H. The Agro-Economic Feasibility of Growing the Medicinal Plant *Euphorbia peplus* in a Modified Vertical Hydroponic Shipping Container. *Horticulturae* **2022**, 8, 256. https://doi.org/10.3390/horticulturae8030256
8. Baily, G.E. Vertical Farming. *Cornell University Library*, **1852**. https://archive.org/details/cu31924000349328
9. Balliu, A.; Zheng, Y.; Sallaku, G.; Fernández, J.A.; Gruda, N.S.; Tuzel, Y. Environmental and Cultivation Factors Affect the Morphology, Architecture and Performance of Root Systems in Soilless Grown Plants. *Horticulturae* **2021**, 7, 243. https://doi.org/10.3390/horticulturae7080243
10. Banerjee C. and Adenaeuer L. Up, Up and Away! The Economics of Vertical Farming, *J. Agric. Stud.* **2014**, vol. 2, no. 1, p. 40. http://large.stanford.edu/courses/2016/ph240/swafford2/docs/banerjee.pdf
11. Barathi, R. D., & Vidjeapriya, R. Life Cycle Cost Analysis of Rooftop Gardens Using OpenLCA. IOP *Conference Series: Earth and Environmental Science* **2022**, 1086(1), 012006. https://doi.org/10.1088/1755-1315/1086/1/012006
12. Beacham, A. M., Vickers, L. H., & Monaghan, J. M. (2019). Vertical farming: a summary of approaches to growing skywards. *The Journal of Horticultural Science and Biotechnology* **2019**, 94, 277-283. https://doi.org/10.1080/14620316.2019.1574214
13. Birlanga, V.; Acosta-Motos, J.R.; Pérez-Pérez, J.M. Mitigation of Calcium-Related Disorders in Soilless Production Systems. *Agronomy* **2022**, 12, 644. https://doi.org/10.3390/agronomy12030644
14. Bisiani, T., Basso, S., Martolana, P., & Venudo, A. (2022). Vertical Farm: from Agriculture to a New City Architecture. Forum A+P **2022**, October. University of Trieste.

https://press.universitetipolis.edu.al/wp-content/uploads/2023/07/5.-Vertical-Farm-Bisiani-et-al.pdf

15. Borrero, J.D. Expanding the Level of Technological Readiness for a Low-Cost Vertical Hydroponic System. *Inventions* **2021**, *6*, 68. https://doi.org/10.3390/inventions6040068
16. Carpineti, C., Meinen, E., Vanacore, L., Leman, A., Barbagli, T., Ketel, E., van Hoogdalem, M., & Janse, J. The added value of indoor products: the strawberry case. *Wageningen University & Research* **2022**. https://doi.org/10.18174/657739
17. Carotti, L., Pistillo, A., Zauli, I., Meneghello, D., Martin, M., Pennisi, G., Gianquinto, G., & Orsini, F. Improving water use efficiency in vertical farming: Effects of growing systems, far-red radiation and planting density on lettuce cultivation. *Agricultural Water Management* **2023**, 285, 108365. https://doi.org/10.1016/j.agwat.2023.108365
18. Carreño-Ortega, A.; do Paço, T.A.; Díaz-Pérez, M.; Gómez-Galán, M. Lettuce Production under Mini-PV Modules Arranged in Patterned Designs. *Agronomy* **2021**, *11*, 2554. https://doi.org/10.3390/agronomy11122554
19. Chakraborty A, Das S, Mondal B. Integrating Neural Network for Pest Detection in Controlled Environment Vertical Farm. Indian Journal of Science and Technology **2022**, 15(17): 829-838. https://doi.org/10.17485/IJST/v15i17.353
20. Chiaranunt P, White JF. Plant Beneficial Bacteria and Their Potential Applications in Vertical Farming Systems. *Plants* **2023** Jan 15;12(2):400. https://doi.org/10.3390/plants12020400
21. Cicekli, M., Barlas N. T. Transformation of today greenhouses into high technology vertical farming systems for metropolitan regions, *J. Environ. Prot. Ecol.*, **2014**. vol. 15, no. 4, pp. 1779–1785
22. Cossu, M., Tiloca, M. T., Cossu, A., Deligios, P. A., Pala, T., & Ledda, L. Increasing the agricultural sustainability of closed agrivoltaic systems with the integration of vertical farming: A case study on baby-leaf lettuce. *Applied Energy* **2023**, Volume 344, 121278. https://doi.org/10.1016/j.apenergy.2023.121278
23. Coutand C, Adam B, Ploquin S, Moulia B. A method for the quantification of phototropic and gravitropic sensitivities of plants combining an original experimental device with model-assisted phenotyping: Exploratory test of the method on three hardwood tree species. *PLoS One.* **2019** Jan 25;14(1):e0209973. https://doi.org/10.1371/journal.pone.0209973
24. Daneshyar, E. Residential Rooftop Urban Agriculture: Architectural Design Recommendations. *Sustainability* **2024**, *16*, 1881. https://doi.org/10.3390/su16051881
25. de Carbonnel, M.; Stormonth-Darling, J.M.; Liu, W.; Kuziak, D.; Jones, M.A. Realising the Environmental Potential of Vertical Farming Systems through Advances in Plant Photobiology. *Biology* **2022**, *11*, 922. https://doi.org/10.3390/biology11060922
26. Dingley, A.; Anwar, S.; Kristiansen, P.; Warwick, N.W.M.; Wang, C.-H.; Sindel, B.M.; Cazzonelli, C.I. Precision Pollination Strategies for Advancing Horticultural Tomato Crop Production. *Agronomy* **2022**, *12*, 518. https://doi.org/10.3390/agronomy12020518
27. Dhawi, F. The Role of Plant Growth-Promoting Microorganisms (PGPMs) and Their Feasibility in Hydroponics and Vertical Farming. *Metabolites* **2023**, *13*, 247. https://doi.org/10.3390/metabo13020247

28. Dou, H., Niu, G., Gu, M., & Masabni, J. Morphological and Physiological Responses in Basil and Brassica Species to Different Proportions of Red, Blue, and Green Wavelengths in Indoor Vertical Farming. *Journal of the American Society for Horticultural Science J. Amer. Soc. Hort. Sci.*, **2020**. *145*(4), 267-278. https://doi.org/10.21273/JASHS04927-20
29. FAO. Global Agriculture Towards 2050. *High-Level Expert Forum*, Rome **2009**. https://www.fao.org/fileadmin/templates/wsfs/docs/Issues_papers/HLEF2050_Global_Agriculture.pdf
30. Farhangi, H., Mozafari, V., Roosta, H.R. et al. Optimizing growth conditions in vertical farming: enhancing lettuce and basil cultivation through the application of the Taguchi method. *Sci Rep* **2023** 13, 6717. https://doi.org/10.1038/s41598-023-33855-z
31. Fernie, A. R., & Yan, J. Targeting Key Genes to Tailor Old and New Crops for a Greener Agriculture. *Molecular Plant* **2020**, 13(3), 354-356. https://doi.org/10.1016/j.molp.2020.02.007
32. Fitzner M., Schreiner M., Baldermann S., Comprehensive characterization of selected phytochemicals and minerals of selected edible halophytes grown in saline indoor farming for future food production, *Journal of Food Composition and Analysis* **2023**, Volume 122, 105435. https://doi.org/10.1016/j.jfca.2023.105435.
33. Franchetti, B.; Ntouskos, V.; Giuliani, P.; Herman, T.; Barnes, L.; Pirri, F. Vision Based Modeling of Plants Phenotyping in Vertical Farming under Artificial Lighting. *Sensors* **2019**, *19*, 4378. https://doi.org/10.3390/s19204378
34. Gao D, Ji X, Pei W, Zhang X, Li F, Han Q, Zhang S. Thermal management and energy efficiency analysis of planar-array LED water-cooling luminaires in vertical farming systems for saffron. *Case Studies in Thermal Engineering*. **2023**, 51:103535. https://doi.org/10.1016/j.csite.2023.103535
35. Ghazal, I.; Mansour, R.; Davidová, M. AGRI|gen: Analysis and Design of a Parametric Modular System for Vertical Urban Agriculture. *Sustainability* **2023**, *15*, 5284. https://doi.org/10.3390/su15065284
36. Gupta, A., Sharma, T., Singh, S. P., Bhardwaj, A., Srivastava, D., & Kumar, R. Prospects of microgreens as budding living functional food: Breeding and biofortification through OMICS and other approaches for nutritional security. Frontiers in Genetics **2023**, 14, 1053810. https://doi.org/10.3389/fgene.2023.1053810
37. Hahm, S.; Lee, B.; Bok, G.; Kim, S.; Park, J. Diniconazole Promotes the Yield of Female Hemp (*Cannabis sativa*) Inflorescence and Cannabinoids in a Vertical Farming System. *Agronomy* **2023**, *13*, 1497. https://doi.org/10.3390/agronomy13061497
38. He, J. Enhancing Productivity and Improving Nutritional Quality of Subtropical and Temperate Leafy Vegetables in Tropical Greenhouses and Indoor Farming Systems. *Horticulturae* **2024**, *10*, 306. https://doi.org/10.3390/horticulturae10030306
39. He, R, Ju, J, Liu, K, Song, J, Zhang, S, Zhang, M, Hu, Y, Liu, X, Li, Y and Liu, H (2024) Technology of plant factory for vegetable crop speed breeding. Front. Plant Sci. 15:1414860. doi: https://doi.org/10.3389/fpls.2024.1414860
40. Henry, C., John, G.P., Pan, R. et al. A stomatal safety-efficiency trade-off constrains responses to leaf dehydration. *Nat Commun* **2019**, 10, 3398. https://doi.org/10.1038/s41467-019-11006-1
41. Hwang, Y.; Lee, S.; Kim, T.; Baik, K.; Choi, Y. Crop Growth Monitoring System in Vertical Farms Based on Region-of-Interest Prediction. *Agriculture* **2022**, *12*, 656. https://doi.org/10.3390/agriculture12050656

42. Ikiz, B., Dasgan, H.Y. & Gruda, N.S. Utilizing the power of plant growth promoting rhizobacteria on reducing mineral fertilizer, improved yield, and nutritional quality of Batavia lettuce in a floating culture. *Sci Rep* **2024** 14, 1616. https://doi.org/10.1038/s41598-024-51818-w
43. Ikiz B, Dasgan HY, Oz BC. Mitigating Tipburn Through Foliar Calcium Application in Indoor Hydroponically Grown Mini Cos Lettuce. *BIO Web Conf.* **2024**. 85:01003. DOI: https://doi.org/10.1051/bioconf/20248501003
44. Isakovic, H., Fasching, A., Punzenberger, L., & Grosu, R. (2019). CPS/IoT Ecosystem: Indoor Vertical Farming System. IEEE 23rd International Symposium on Consumer Technologies (ISCT) 2019, 47-52. Retrieved from https://api.semanticscholar.org/CorpusID:208038293
45. Jung, A., Szabó, D., Varga, Z., Pék, Z., Vohland, M., & Sipos, L. Spatially scaled and customised daily light integral maps for horticulture lighting design. *NJAS: Impact in Agricultural and Life Sciences* **2024**, *96*(1). https://doi.org/10.1080/27685241.2024.2349522
46. Jurga, A.; Pacak, A.; Pandelidis, D.; Kaźmierczak, B. A Long-Term Analysis of the Possibility of Water Recovery for Hydroponic Lettuce Irrigation in an Indoor Vertical Farm. Part 2: Rainwater Harvesting. *Appl. Sci.* **2021**, *11*, 310. https://doi.org/10.3390/app11010310
47. Kabir, M.S.N.; Reza, M.N.; Chowdhury, M.; Ali, M.; Samsuzzaman; Ali, M.R.; Lee, K.Y.; Chung, S.-O. Technological Trends and Engineering Issues on Vertical Farms: A Review. *Horticulturae* **2023**, *9*, 1229. https://doi.org/10.3390/horticulturae9111229
48. Kalantari, F., Kalantari, S., Deshkar, A., Hattimare, N., & Jadhav, S. Determination of Optimal Daily Light Integral (DLI) for Indoor Cultivation of Iceberg Lettuce in an Indigenous Vertical Hydroponic System. *Scientific Reports* **2023**, 13, 36997. https://doi.org/10.1038/s41598-023-36997-2
49. Kamenchuk, V.; Rumiantsev, B.; Dzhatdoeva, S.; Sadykhov, E.; Kochkarov, A. Analysis of Cross-Influence of Microclimate, Lighting, and Soil Parameters in the Vertical Farm. *Agronomy* **2023**, 13, 2174. https://doi.org/10.3390/agronomy13082174
50. Kelada, K.D.; Tusé, D.; Gleba, Y.; McDonald, K.A.; Nandi, S. Process Simulation and Techno-Economic Analysis of Large-Scale Bioproduction of Sweet Protein Thaumatin II. *Foods* **2021**, *10*, 838. https://doi.org/10.3390/foods10040838
51. Khan, N., Bano, A., Ali, S. et al. Crosstalk amongst phytohormones from planta and PGPR under biotic and abiotic stresses. *Plant Growth Regul* **2020**. 90, 189–203. https://doi.org/10.1007/s10725-020-00571-x
52. Lee, JY, Rahman, A, Azam, H, Kim, HS, Kwon, MJ. Characterizing nutrient uptake kinetics for efficient crop production during *Solanum lycopersicum var. cerasiforme* Alef. growth in a closed indoor hydroponic system. PLoS ONE **2017**, 12(5): e0177041. https://doi.org/10.1371/journal.pone.0177041
53. Maluin, F.N.; Hussein, M.Z.; Nik Ibrahim, N.N.L.; Wayayok, A.; Hashim, N. Some Emerging Opportunities of Nanotechnology Development for Soilless and Microgreen Farming. *Agronomy* **2021**, *11*, 1213. https://doi.org/10.3390/agronomy11061213
54. Min, K.; Lim, D. Designing Automated Logistics Warehouse Stackable Bidirectional Infinite-Loop Modules. *Appl. Sci.* **2023**, *13*, 12472. https://doi.org/10.3390/app132212472

55. Mishra, S., Khoda, K., Yau, Y.-Y., & Easterling, M. Vertical Cultivation: Moving Towards a Sustainable and Eco-friendly Farming. *Springer-Nature* **2021**. https://doi.org/10.1007/978-981-16-9001-3_20
56. Gentry, M. Local heat, local food: Integrating vertical hydroponic farming with district heating in Sweden, Energy, Elsevier **2019**, vol. 174(C), pages 191-197.
57. Lastochkina, O.; Aliniaeifard, S.; SeifiKalhor, M.; Bosacchi, M.; Maslennikova, D.; Lubyanova, A. Novel Approaches for Sustainable Horticultural Crop Production: Advances and Prospects. *Horticulturae* **2022**, *8*, 910. https://doi.org/10.3390/horticulturae8100910
58. Lee, S.; Song, M.-J.; Oh, M.-M. Effects of Air Anions on Growth and Economic Feasibility of Lettuce: A Plant Factory Experiment Approach. *Sustainability* **2022**, *14*, 15468. https://doi.org/10.3390/su142215468
59. Lubna, F.A.; Lewus, D.C.; Shelford, T.J.; Both, A.-J. What You May Not Realize about Vertical Farming. *Horticulturae* **2022**, *8*, 322. https://doi.org/10.3390/horticulturae8040322
60. Mahalingam D, Patankar A, Phi K, Chakraborty N, McGann R, Ramakrishnan IV. Containerized Vertical Farming Using Cobots. *arXiv*. **2023**. Available from: https://doi.org/10.48550/arXiv.2310.15385
61. Moghimi F, Asiabanpour B. Economics of Vertical Farming: Quantitative Decision Model and a Case Study for Different Markets in the USA. *Research Square.* **2021**. https://doi.org/10.21203/rs.3.rs-943119/v1
62. Mustapha, S., Musa, A. K., Apalowo, O. A., Lawal, A. A., Olayiwola, O. I., Bamidele, H. O., & Uddin II, R. O. Open vertical farms: a plausible system in increasing tomato yield and encouraging natural suppression of whiteflies. Acta Agriculturae Slovenica **2022**, 118(2), 1–9. https://doi.org/10.14720/aas.2022.118.2.2272
63. Nájera, C.; Gallegos-Cedillo, V.M.; Ros, M.; Pascual, J.A. Role of Spectrum-Light on Productivity, and Plant Quality over Vertical Farming Systems: Bibliometric Analysis. *Horticulturae* **2023**, *9*, 63. https://doi.org/10.3390/horticulturae9010063
64. Naranjani, B., Najafianashrafi, Z., Pascual, C., Agulto, I., & Chuang, P.-Y. A. Computational analysis of the environment in an indoor vertical farming system. *International Journal of Heat and Mass Transfer* **2022**, Volume 186, 122460. https://doi.org/10.1016/j.ijheatmasstransfer.2021.122460.
65. Nguyen, T.K.L.; Cho, K.M.; Lee, H.-Y.; Sim, H.-S.; Kim, J.-H.; Son, K.-H. Growth, Fruit Yield, and Bioactive Compounds of Cherry Tomato in Response to Specific White-Based Full-Spectrum Supplemental LED Lighting. *Horticulturae* **2022**, *8*, 319. https://doi.org/10.3390/horticulturae8040319
66. Peterswald TJ, Mieog JC, Azman Halimi R, Magner NJ, Trebilco A, Kretzschmar T, Purdy SJ. Moving Away from 12:12; the Effect of Different Photoperiods on Biomass Yield and Cannabinoids in Medicinal Cannabis. *Plants* **2023**;12(5):1061. https://doi.org/10.3390/plants12051061
67. Ptak, M.; Wasieńko, S.; Makuła, P. A New Approach for Vertical Plant Cultivation Maximizing Crop Efficiency. Preprints **2024**, 2024061047. https://doi.org/10.20944/preprints202406.1047.v1
68. Rathor, A. S., Choudhury, S., Sharma, A., & Nautiyal, P. Empowering Vertical Farming through IoT and AI-Driven Technologies: A Comprehensive Review. *Heliyon* **2024**, 10, e34998. https://doi.org/10.1016/j.heliyon.2024.e34998
69. Righini, I., Graamans, L., van Hoogdalem, M., Carpineti, C., Hageraats, S., van Munnen, D., Elings, A., de Jong, R., Wang, S., Meinen, E., Stanghellini, C., Hemming, S. and Marcelis, L.F., Protein plant factories: production and resource use efficiency of soybean proteins in vertical farming. *J Sci Food Agric* **2024**, 104: 6252-6261. https://doi.org/10.1002/jsfa.13458

70. Roberts, J. Vertical Farming Systems Bring New Considerations for Pest and Disease Management. *Horticultural Science Journal*, **2020**, Issue 2. https://doi.org/10.1016/j.hortsci.2020.02.001
71. Runke, E. Daily Light Integral Defined. **2006**. Michigan State University Extension. Retrieved from https://www.canr.msu.edu/resources/daily_light_integral_defined
72. Runke, E. Do you know what your DLI is? **2006**. Michigan State University Extension. Retrieved from https://www.canr.msu.edu/resources/do_you_know_what_your_dli_is
73. Saad, M.H.M.; Hamdan, N.M.; Sarker, M.R. State of the Art of Urban Smart Vertical Farming Automation System: Advanced Topologies, Issues and Recommendations. *Electronics* **2021**, *10*, 1422. https://doi.org/10.3390/electronics10121422
74. Sandison, F., Yeluripati, J., & Stewart, D. Does green vertical farming offer a sustainable alternative to conventional methods of production?: A case study from Scotland. *Food and Energy Security* **2023**, 12, e438. https://doi.org/10.1002/fes3.438
75. Scavo, A. J., Sidhom, M., & Rangel, F. J. Possible impacts of rising CO2 on crop water use efficiency and food security. 2018. Retrieved from https://escholarship.org/content/qt73n3p64x/qt73n3p64x.pdf?t=ph0cdy
76. Sharath Kumar M., Luo J., Xi Y., van Ieperen W., Marcelis L.F.M., Heuvelink E., Several short-day species can flower under blue-extended long days, but this response is not universal, *Scientia Horticulturae* **2024**, 325, 112657, https://doi.org/10.1016/j.scienta.2023.112657
77. Sharma, A., Hazarika, M., Heisnam, P., Pandey, H., Nampoothiri Devadas, V. A. S., Kesavan, A. K., Kumar, P., Singh, D., Vashishth, A., Jha, R., Misra, V., & Kumar, R. Controlled environment ecosystem: A cutting-edge technology in speed breeding. *ACS Omega* **2024**, 9(27), 29114-29138. https://doi.org/10.1021/acsomega.3c09060
78. Sharma, R. K., Chauhan, V. S., Gupta, S. D., Patel, P. R. An IoT Based Drip Irrigation System for Vertical Farming in Rainshelter. *Journal of Agricultural Technology*, **2023**, Issue 3. https://doi.org/10.1016/j.jagtec.2023.05.012
79. Schmidt Rivera X, Rodgers B, Odanye T, Jalil-Vega F, Farmer J. The role of aeroponic container farms in sustainable food systems – The environmental credentials. *Science of The Total Environment*. **2023**. 860:160420. https://doi.org/10.1016/j.scitotenv.2022.160420
80. Siregar, R.R.A.; Seminar, K.B.; Wahjuni, S.; Santosa, E. Vertical Farming Perspectives in Support of Precision Agriculture Using Artificial Intelligence: A Review. *Computers* **2022**, *11*, 135. https://doi.org/10.3390/computers11090135
81. Soh, C. B., Haridarshan, R., Chien, S.-C., An, H., Saha, A., & Teoh, M. T. Study on Outdoor Urban Farming Planters Through Daylight Simulations: A Full-Scale Experiment in Singapore. *14th Asia Lighting Conference Proceedings* **2023**. https://irr.singaporetech.edu.sg/articles/conference_contribution/Study_on_Outdoor_Urban_Farming_Planters_Through_Daylight_Simulations_A_Full-Scale_Experiment_In_Singapore/24224599
82. Song, S., Ong, E. J. K., Lee, A. M. J., & Chew, F. T. How crop breeding programs can improve plant factories' business and environmental sustainability: Insights from a farm level analysis. *Sustainable Production and Consumption* **2024**, 44, 298-311. https://doi.org/10.1016/j.spc.2023.12.020
83. Stupariu IM, Balint MV, Poșta Gh. Research on the influence of some technological factors on the culture of microgreens in vertical farming. *Journal of Horticulture, Forestry and Biotechnology*. **2023**; 27(3):135-140. http://www.journal-hfb.usab-tm.ro

84. Traykova B.D., Stanilova M.I., Soilless Propagation of *Haberlea rhodopensis Friv.* Using Different Hydroponic Systems and Substrata, Ecologia Balkanica **2020**, Vol. 12, Issue 1, pp. 111-121.
85. Twalla, J.T., Ding, B., Cao, G. et al. Roles of stomata in gramineous crops growth and biomass production. *Cereal Research Communications* **2022** 50, 603–616 (2022). https://doi.org/10.1007/s42976-021-00216-3
86. van Delden, S.H., Sharath Kumar, M., Butturini, M. et al. Current status and future challenges in implementing and upscaling vertical farming systems. *Nat Food* **2021**, 2, 944–956. https://doi.org/10.1038/s43016-021-00402-w
87. Van Gerrewey, T.; Boon, N.; Geelen, D. Vertical Farming: The Only Way Is Up? *Agronomy* **2022**, *12*, 2. https://doi.org/10.3390/agronomy12010002
88. Vatistas, C.; Avgoustaki, D.D.; Bartzanas, T. A Systematic Literature Review on Controlled-Environment Agriculture: How Vertical Farms and Greenhouses Can Influence the Sustainability and Footprint of Urban Microclimate with Local Food Production. *Atmosphere* **2022**, *13*, 1258. https://doi.org/10.3390/atmos13081258
89. Wichitwechkarn, V., Rohde, W., & Choudhary, R. (2023). Design and validation of an open-sourced automation system for vertical farming. HardwareX, 16, e00497. https://doi.org/10.1016/j.ohx.2023.e00497
90. Williams, K., Subramani, M., Lofton, L. W., Penney, M., Todd, A., & Ozbay, G. Tools and Techniques to Accelerate Crop Breeding. *Plants* **2024**, 13(1520). https://doi.org/10.3390/plants13111520
91. Wong, C. E., Teo, Z. W. N., Shen, L., & Yu, H. Seeing the lights for leafy greens in indoor vertical farming. *Trends in Food Science & Technology* **2020**, 106, 48-63. https://doi.org/10.1016/j.tifs.2020.09.031
92. Xu, J., Guo, Z., Jiang, X., Ahammed, G. J., & Zhou, Y. (2021). Light regulation of horticultural crop nutrient uptake and utilization. *Horticultural Plant Journal* **2021**, 7(5), 367-379. https://doi.org/10.1016/j.hpj.2021.01.005
93. Xu, W.; Lu, N.; Kikuchi, M.; Takagaki, M. Continuous Lighting and High Daily Light Integral Enhance Yield and Quality of Mass-Produced Nasturtium (*Tropaeolum majus* L.) in Plant Factories. *Plants* **2021**, *10*, 1203. https://doi.org/10.3390/plants10061203
94. Yan, J., Bogie, N. A., & Ghezzehei, T. A. Root uptake under mismatched distributions of water and nutrients in the root zone. *Biogeosciences* **2020**, 17*(24), 6377–6392. https://doi.org/10.5194/bg-17-6377-2020
95. Yang, J.; Song, J.; Jeong, B.R. Side Lighting Enhances Morphophysiology and Runner Formation by Upregulating Photosynthesis in Strawberry Grown in Controlled Environment. *Agronomy* **2022**, *12*, 24. https://doi.org/10.3390/agronomy12010024
96. Zandalinas, S., Fritschi, F., Mittler, R. Global Warming, Climate Change, and Environmental Pollution: Recipe for a Multifactorial Stress Combination Disaster. *Trends in Plant Science* **2021**. 26 (6): 588–599. https://doi.org/10.1016/j.tplants.2021.02.011
97. Zhang, X., Kuru, A., Brambilla, A., & Gasparri, E. Potential and Challenges of Vertical Farming on Building Facades in Practice. *PLEA 2024 Conference Proceedings* **2024**. https://www.researchgate.net/publication/382109743

About the Author

Dr. Maryna Kuzmenko is a Ukraine-born but UK-based serial entrepreneur focused on agriculture technologies, education, and sustainability. Maryna possesses a wealth of experience in business and operations, honed and refined at her AI platform, Petiole Pro, for plant phenotyping and quality assurance using a smartphone.

She holds a Ph.D. in Business Law from one of Ukraine's leading universities and the IEMA Foundation Certificate in Environmental Management.

She is deeply passionate about urban farming and owns two micro-urban farming setups: one on her windowsill and another in her backyard.

Maryna is always delighted to know more about her readers.
Drop an email to maryna@petioleapp.com or connect with her on Linkedin ♥

QR code of Maryna's Linkedin

About the Publisher

Colophon

The cover of *Vertical Farming: A Guide for Growing Minds* features a stylized botanical portrait of **Albert Einstein** (1879-1955), symbolising the innovative spirit of vertical farming.

Einstein, like vertical farming, was initially misunderstood by his contemporaries but eventually recognized for his groundbreaking ideas.

This parallel reflects the journey of vertical farming itself. Once met with scepticism, this innovative agricultural method is now gaining acclaim for its potential to revolutionise food production.

Just as Einstein's theories transformed our understanding of the universe, vertical farming is reshaping our approach to sustainable agriculture.

www.ingramcontent.com/pod-product-compliance
Lightning Source LLC
Chambersburg PA
CBHW031610210526
45464CB00004B/1509